MW00454697

Documenting Aftermath

Infrastructures Series

edited by Geoffrey C. Bowker and Paul N. Edwards

Documenting Aftermath

Information Infrastructures in the Wake of Disasters

Megan Finn

The MIT Press
Cambridge, Massachusetts
London, England

This book was set in ITC Stone Serif Std by Toppan Best-set Premedia Limited. Printed and bound in the United States of America.

Library of Congress Cataloging-in-Publication Data

Names: Finn, Megan, author.
Title: Documenting aftermath : information infrastructures in the wake of disasters / Megan Finn.
Description: Cambridge, MA : MIT Press, [2018] | Series: Infrastructures | Includes bibliographical references and index.
Identifiers: LCCN 2017056201 | ISBN 9780262038218 (hardcover : alk. paper)
Subjects: LCSH: Disaster relief--Social aspects--United States--History--20th century. | Earthquakes--California--Press coverage--History--20th century. | Information policy--California--History--20th century. | Emergency management--California--History--20th century. | Mass media--California--History--20th century.
Classification: LCC HV555.U6 F56 2018 | DDC 363.34/80973--dc23 LC record available at https://lccn.loc.gov/2017056201

10 9 8 7 6 5 4 3 2 1

For Kenn

Contents

Contents

Preface: Toward an Informatics of Disaster

> It's become a ritual that when earthquakes hit Tokyo, people announce on Twitter as soon as they feel them. So I opened up Twitter and sure enough, pages of "Oh, felt that," "Shaking," "Earthquake!" filled my screen. Then things got weird.[1]

As the horrific impacts of the quake began to sink in, Tokyo resident Steve Nagata set up a video camera, pointed it at his TV, and began live streaming the news broadcasts. Over the next few days, hundreds of people desperately seeking news and information about the disaster began to follow his accounts:

> Together we learned of the tsunami and the terrible damage, the towns washed away into the ocean. Together we learned of the problems with the nuclear reactors in Fukushima. Together we learned that the entire world had heard of the disaster and that planes were on their way to help those in need. Together with thousands of people in my online community, most of whom I have never met, I felt fear, gratitude and sometimes despair, but I never felt alone.

Nagata's experience was not unique. The Internet was integrated into how people near and far experienced the 2011 Tōhoku earthquake and tsunami in Japan, which killed over fifteen thousand people. The information and communication infrastructures not only allowed people to understand what had happened but also shaped possibilities for action.

Brent Stirling, once a resident of Fukushima City who lived in Ottawa, Canada during the earthquake, remembered, "I got on Skype and contacted every single person I knew in Japan that I could and asked about their situation." He tried to pass on helpful updates to people who were in the area: "From NHK and BBC news, I posted Facebook and Twitter updates for the people in Fukushima about the status of gasoline, trains, roads, and the weather. ... It was incredible to watch how current and former residents of Fukushima helped each other get the information they needed." People far away from the earthquake experienced it through the news stations (both on television and online), social media platforms, and Internet-mediated interactions with other people struggling to comprehend an incomprehensible event.

This arrangement of technologies, people, and institutions also shaped the actions of those who felt the shaking. One Tokyo resident told of being in an underground subway when the earthquake hit. On getting out to the street, he immediately saw a crowded park filled with people "frantically typing on their mobile phones." Many of the stories that emerged in the aftermath, particularly from those who felt the earthquake but were not in the tsunami's path, involved tales of first securing their own safety and that of the people around them, and then turning to a variety of media sources to learn about what had happened, tell others about their well-being, and find out about loved ones they had not heard from.[2] Edan Corkill recalled being at work and shaking as he "bashed out another email" to his wife who was at their home by the sea in the tsunami warning zone, beseeching her to "climb the mountain. ... I tried again and again to call my wife, but it was hopeless." For Joel David Neff, "[My] first thought was to update my status on Facebook."

These remembrances come from a collection of essays and other artwork called *2:46: Aftershocks: Stories from the Japan Earthquake* that describes the enormity of what is now known as the 2011 Tōhoku earthquake and tsunami. The first-person accounts tell the stories of people's experiences during the earthquake and in the days following it.[3] Many of the book's stories place information and communication practices at the center of their narratives, but *2:46* itself is also a representation of modern public information infrastructures.[4] The volunteers who contributed reflections to the book found each other using the Twitter hashtag #quakebook. In fact, a draft of the entire manuscript was completed in a single week, and the whole book was finished a mere two weeks after the earthquake.

And yet although information infrastructure functioned well enough to assemble *2:46*, the collection is also littered with examples of infrastructure failure. In times of disaster, the nonworking infrastructure is often more noteworthy, serving as another source for interpreting the impacts of the disaster. After the earthquake in Japan, people imagined the worst when they did not hear from their loved ones. In Hamilton, Canada, Kevin Wood and his family waited for news from their family in Miyagi. "Hours of anxiety stretched into days," until "we called, and at long last, they answered." Restoring services was no easy feat, requiring significant labor and resources in the aftermath of a major disaster. Kosuke Ishihara worked for one of the telecommunication companies in the hardest-hit areas and reflected on the work of the support crews: "These workers had family members that were missing, yet they threw themselves into their work." In addition to using telephones and the Internet to try to reach their own friends and family, telecommunication workers found themselves as key actors in ensuring the continuity of a working infrastructure.

Besides contacting loved ones, people turned to their Internet-enabled devices to learn more generally about what had happened and understand how to be safe. This was salient for earthquake survivors living near damaged nuclear reactors and relying on the Tokyo Electric Power Company for updates about radiation risk.[5] Some of the stories in *2:46* reflect this reliance on private and public institutions for information, even though this was also coupled with a distrust of many authorities. People felt that different media sources sensationalized or downplayed danger, and were frustrated with a lack of clear information about whether those living close to the nuclear reactor were in imminent danger.

One narrative from the collection, from Grandfather Hibiki, stands out because it describes an information experience of someone not online. Hibiki was in an area without power, and relied on the radio for news and information: "Many of us oldies are familiar with the radio and listen to late-night broadcasts, with batteries that last a surprisingly long time." But Internet surrogates helped these "oldies" get additional information beyond the radio broadcasts. "We can ask net-savvy people living near us to get this information for us." Although Hibiki didn't access the Internet himself, his information experience was still mediated by digital technology. Thus, as we marvel at the speed of production of *2:46*, and even suppose that the new infrastructural technologies precipitate new ways of knowing, the story of Hibiki makes clear that a description of postdisaster information practices in Japan focused solely on how the earthquake was mediated by the Internet would be incomplete because of the myriad of overlapping and discrete means for circulating information.

The stories in *2:46* describe who got what information when, where people expected it to come from, what they believed it should tell them, and what was considered informative, as people got in touch with loved ones, tried to figure out what happened, and determined what to do next. The multilayered information order of *2:46* includes assemblages of media and technology companies, government organizations, physical telecommunication infrastructure, different devices, Internet platforms, television stations, radios, and a multitude of information practices that facilitated the production and circulation of information for publics. The variety of different devices, people using and not using them, telecommunication workers repairing the physical infrastructures, and power companies disseminating information all shaped how people across Japan as well as around the world experienced the disaster and its aftermath. These stories about the 2011 Tōhoku earthquake and tsunami underscore the relevance of information practices after disasters, specific to a certain time and place.

In this book, I move across the Pacific Ocean from Japan and examine information orders after earthquakes in Northern California in 1868, 1906, and 1989, as well as the potential information order in place today. *Documenting Aftermath* explores the different informational entities, from the post office and Red Cross to the Federal Emergency Management Agency (FEMA) and Twitter, that produce and circulate information for various publics, and how these information streams shape people's experiences of disaster in historical and contemporary Northern California.

Acknowledgments

The California State Archives, the California State Libraries, San Francisco Public Library, California Historical Society, the National Archives in College Park and Menlo Park, the Society of California Pioneers, Stanford's Jonsson Library of Government Documents, and the University of California Libraries were key research sites for me. A special thanks to many talented librarians and archivists at the University of California at Berkeley, especially at the Bancroft Library, Interlibrary Loan, Institute for Government Study, and Doe Library. These people and institutions connected me with the documents that made the stories in the following pages possible.

On my bookshelf sits a copy of *Sorting Things Out* that belonged to my late master's thesis adviser, Peter Lyman, with a simple inscription by Leigh Star: "fellow traveler." I thank my fellow travelers: Daniela Rosner, Mike Ananny, Christo Sims, Ashwin Mathew, Morgan Ames, Janaki Srinivasan, Dan Perkel, Elisa Oreglia, Kate Miltner, Jessa Lingel, Amelia Acker, Lilly Irani, Sareeta Amrute, Lilly Nguyen, Christina Dunbar-Hester, Sarah Quinn, danah boyd, Alice Marwick, Rajesh Veeraraghavan, T. L. Taylor, Teddy Chao, Holly Jones, and Julie Burrelle.

This project started as a dissertation at Berkeley's School of Information, in my experience a community with a foundation of intellectual freedom, scholarly rigor, and relationships based on respect and kindness. I'm grateful for the support of all the faculty and staff, particularly advisers and teachers AnnaLee Saxenian, Coye Cheshire, Michael Buckland, John Chuang, and Peter Lyman. I owe a huge debt to Paul Duguid for his brilliant and gentle guidances.

This project became a book with the support of the Social Media Collective at Microsoft Research New England. Mary Gray and Nancy Baym helped me throughout my postdoc and beyond. I am also grateful for the support that I have received at University of Washington's Information School.

I thank Margy Avery, Robin DeCook, and Laura Portwood-Stacer for their thoughtful editorial guidance on this book. At the MIT Press, Katie Helke and Deborah Cantor-Adams have helped actualize this book. I'm deeply appreciative of Paul Edwards's and Geoffrey Bowker's work on the Infrastructures series.

I received valuable comments on book sections from colleagues at the History of Data workshop at Columbia, Knowledge Infrastructures workshop at University of Washington, Society for the History of Technology SIGCIS works-in-progress, Infrastructure Studies reading group at the University of Washington, and three anonymous manuscript reviewers.

Terry, Kris, Claire, Andrew, Justin, Everett, June, and Noelle make sure that I don't take myself too seriously.

I'm so lucky to have Vivi, Terrence, and Kenn in my life, and we were thrilled to welcome Terrence to the family as I finished the final book edits. Vivi's joy, tenacity, and curiosity remind me why I took on this project. Kenn's love and generosity made it possible for me to finish the book.

The work for this book was supported by an NSF Dissertation Improvement Grant.

1   Making Sense of Earthquakes: Public Information Infrastructures and Postdisaster Event Epistemologies

Earthquakes are unique moments of uncertainty where the very ground we walk on ceases to be reliable, and lives, livelihoods, and possessions can be instantly destroyed. Information and communication technologies moderate our experiences after these fraught moments, and can themselves become the objects of intense attention. If an earthquake happened in the San Francisco Bay Area today, many people would expect that the US Geological Survey (USGS) would quickly provide an online map showing the earthquake epicenter, magnitude, and reach. People might hope that the Federal Emergency Management Agency (FEMA), a unit of the Department of Homeland Security, along with branches of local government, would help those affected, and that government officials would get on Twitter or television to tell people where to go for aid and how to be safe. Those impacted by a quake would want to assure loved ones of their well-being; some people would attempt to do this with a phone call, and many would notify their loved ones via Facebook.

The concept of an *information order*, introduced by the late historian C. A. Bayly, points to the many different institutions, technologies, and practices that characterize how groups of people produce and share information as well as exercise power and control.[1] The San Francisco Bay Area's contemporary information order is a complex constellation of information, technologies, media organizations, government institutions, and scientific explanations of disasters that shape how people use information to make sense of an earthquake's aftermath. Through this book, I hope to make these pillars of today's information order—e.g., FEMA, USGS, Facebook—feel peculiar. One hundred and fifty years ago, institutions such as FEMA did not exist, and obtaining accurate and timely information after a disaster was a complex endeavor, raising the question: What kinds of information orders existed after earthquakes in the past, and how did they shape disaster experiences?

The primary argument of this book is that information orders influence how we act in, experience, and document an event. Information orders establish how people are informed about events and facilitate particular possibilities for knowledge, which I call *event epistemologies*. The concept of event epistemology helps explains how a disaster can be shaped by an information order and how an information order can be shaped by disaster. Event epistemologies are constituted partially in documents by and for actors in historical moments, and partially by researchers like me today. But these event epistemologies are not the sterile province of academic researchers; they have material consequences for what is at risk, who is harmed, and how aid is distributed.[2]

As researchers today try to sort out how information technologies impact contemporary events such as the 2011 Tōhoku earthquake and tsunami, we wonder if new technical developments promise a wholly novel event epistemology. Some contemporary information practices, such as those that make use of social media platforms, might appear to be novel, while others might seem to have deep roots with a historical context. But is this really the case? How is the materiality of technology connected or not to shifts in practice? When people describe their postdisaster well-being via a social media platform such as Twitter, we can ask if this practice of notification is specific to this platform or this historical moment. Are there information practices that transcend social or historical contexts, or are they always specific to a time and place? To help evaluate these claims of new modes of producing and circulating knowledge, I explore historical disaster information practices. This book examines postearthquake information orders in Northern California in 1868, 1906, and 1989, and the information order in place if there was an earthquake today. Looking at the historical contexts for how people are informed about a disaster allows us to address questions about the relationship between changing technological infrastructures and changing information practices.

Additionally, from a more instrumental point of view, historical cases of disasters offer readers the opportunity to reimagine present technologies, policies, and information infrastructures because the past looks both familiar and foreign. Denaturalizing the contemporary information orders highlights how people have made the present-day institutions, infrastructures, and practices in ways that promote particular interests at the expense of others. With this in mind, we can ask, What kind of information order are we making today, and how might it meet our ideals, or not?

I analyze information orders in postearthquake contexts when physical infrastructure is often broken. Unfortunately, these moments of infrastructure destruction occur at precisely those instances when people most intensely want to learn about what has happened, and find out about the fates of their friends and loved ones. In a disaster context, the information order is both intensely important and being tested. Within each historical and contemporary case in this book, I also consider the question, What happens when parts of the information order are injured or even destroyed? Earthquakes and other traumatic events punctuate everyday information practices and provide new vantage points for examination. In these moments, many of the institutions, infrastructures, and practices that comprise information orders are simultaneously injured and most needed.

Through examining historical and contemporary information orders, this book reflects on the following comparative question: How do different information orders shape event epistemologies? Understanding how information orders facilitate possibilities for knowledge after a disaster has consequences for many: survivors, aid organizations, and researchers.

Information Orders

The concept of an information order points to the many different infrastructures, institutions, technologies, and practices that might characterize how societies produce and share documents. When Bayly used the term, he was studying colonial India, examining both the information practices of the British occupiers and activities of North Indian society. Though he drew an analytic distinction between the British and Indian practices, he noted that the "state" and "autonomous networks of social communication within Indian society" overlapped. He wrote, "In order to bring the analysis of the two dimensions together, we have used the concept of the information order. This has been taken to include the uses of literacy, social communication and the information knowledge and systems of surveillance of the emerging state."[3] For Bayly, the relations among the different information systems and practices constituted a singular information order.

Moreover, information orders, constituted in sociomaterial relations, vary from society to society:

> The information order of different societies cannot either be reduced to material factors or be viewed simply as a discursive property of the power of elites and the state. It was, rather, a relatively autonomous domain constituted not only by state surveillance and elite ideological representations but also by affective communities of religion, belief, kinship, pilgrimage, literary or linguistic sensibility and styles of political debate. Societies at similar levels of economic development—late Mughal India and Tokugawa Japan, for example—had created information orders which differed significantly from each other.[4]

This insight concurs with intuition about different contemporary disasters; outsiders wishing to know what happened in the Tōhoku earthquake and tsunami in 2011, Hurricane Katrina in 2005, or the Myanmar cyclone in 2008 were confronted with vastly different information orders in each historical moment in each geographic area.

The concept of an information order should not be overextended or simplified. Yet the information order as a "heuristic device" is not without power: it is both related to the political and economic orders while being distinct from them.[5] Bayly says that since researchers regularly refer to a society's economic or political order, it also makes sense to discuss how information orders shape a society. The information order is influenced by politics, economic forces, and material artifacts, but it is not wholly determined by these things. Thus we can talk about an information order at different historical moments without it being a totalizing concept. Inspired by Bayly's heuristic, I unpack information orders in different historical moments in Northern California to understand how they shaped event epistemologies after earthquakes.

Infrastructures

One of the key analytics that I develop in this book to examine the information order is that of public information infrastructure. The term *public information infrastructure* builds on an important body of work from a number of disciplines that examines how infrastructures generally, and information infrastructures specifically, shape the production and circulation of information. Public information infrastructures bring together institutions and materials in a particular time and place. How multiple infrastructures articulate together—as in the English and Indian assemblages analyzed by Bayly—is captured by the information order. After an earthquake, public information infrastructures both assist and limit the production and circulation of information about what has happened. In the following sections, I explore the meanings of infrastructure, information and information infrastructure, and public information infrastructure.

Infrastructures are widely understood by scholars to be "pervasive, enabling resources in the networked form."[6] They are the social, economic, and material resources that enable me to believe that when I take a picture and text it to a friend using my smartphone, my friend will receive it. For me to send a single picture of earthquake damage to my friend requires (but is not limited to) a complicated assemblage that includes physical infrastructure such as cell phone devices, phone towers, switches, and cables; standards that allow phones to share and transmit photographs, such as multimedia messaging service and wireless application protocol;

institutions that establish these standards; corporations that pay workers to build and maintain the infrastructure; knowledge that my friend has a phone capable of receiving photographs; and conventions associated with making shared meanings from photographs. Accordingly, infrastructure cannot be simply thought of as the physical tubes and wires but also must include a complex set of organizations, practices, and standards.[7]

But not everyone experiences the same assemblages as infrastructure. Those with access to a cell phone with an expensive data plan might view the telecommunications infrastructure as an enabling resource, while that same enabling quality of infrastructure may not be assumed by those who cannot use a cell phone, or cannot afford a cell phone or data plan that allows them to send pictures. Furthermore, a phone company maintenance employee will not view the cell phone infrastructure as a pervasive enabling resource while at work. At home, though, this employee might understand the assemblage that ensures the delivery of a picture via text message as infrastructure. Based on insights such as these, researchers who study infrastructure from a relational perspective shift the question "What is infrastructure?" to "When is an infrastructure?"—a move that highlights that experiencing something as infrastructure depends on individual status relative to infrastructure.[8] The relational quality of infrastructure highlights differential experiences of it.

Infrastructures are relational and also overlapping.[9] In order to text someone an image, the telecommunications infrastructure relies on a number of additional infrastructures, most obviously electricity, but also the roads that bring workers to their offices, the payment systems that facilitate commerce for the customers, and more; it is these interconnected layers of infrastructures that absorb an earthquake.

Yet despite the challenges to definitions of infrastructure raised by a relational perspective, scholars still ontologically stabilize infrastructure because of the materials, institutions, and practices involved in the continual work of producing infrastructure, even though all people don't experience it as infrastructure. Brian Larkin says that "discussing an infrastructure is a categorical act. It is a moment of tearing into those heterogeneous networks to define which aspect of which network is to be discussed and which parts will be ignored."[10] In this book, my categorical act focuses on information infrastructure.

Information Infrastructures

What, then, is an *information* infrastructure?[11] As a scholar concerned with histories of the intersection of information, technology, and society, I start with the assumption that the idea of information as a commodity is an artifact of late modernity.[12] The term *information* is used in a variety of ways that are quite unspecific and even contradictory.[13] I do not assume consistent usage of information in my sources; I am interested, however, in similar phenomena in different historical moments. Even though information is such an "ill-fitting suit," I use it to mean something close to what Michael Buckland calls *information-as-thing*, or *documents*.[14] A document is intentionally recorded and has a social context; it does not exist naturally in the environment. Documents have a material form, but do not necessarily correspond to any truth claims; documents are merely traces of something. The term document is particularly useful for analyses of communications in historical contexts, because I rely on documents (e.g., newspapers, letters, pictures, and government reports) to provide evidence of and explanations for people's postearthquake actions.

The materiality of documents helps me explain information infrastructure as the ongoing processes involved in the production and circulation of documents.[15] Even though the concept of information is sometimes naturalized, what gets called information must be produced.[16] For example, although a cracked sidewalk after an earthquake could inform someone of a surface fault rupture and therefore be called information, one cannot design an information system that accounts for all cracks in the sidewalk if those cracks in the sidewalk are not represented in some other form. If a photograph of a cracked sidewalk was posted to a news website, or if someone sent a text message to a friend describing the location of the cracked sidewalk, these documents would constitute information as mediated representations.

Canonical works on information infrastructures emphasize that infrastructure is an ongoing process, also called *infrastructuring*.[17] The processual nature of information infrastructure underscores its endurance and continuity.[18] Thus, even though infrastructures are often conceived of as future-shaping projects, they must contend with the past.[19] Much research has supported the idea of infrastructure as sociomaterial and an ongoing process of production, while also emphasizing that when the process stops, infrastructure can quickly degrade.[20] This continual production of information infrastructure simultaneously enables and constrains action.[21] People make the information infrastructures that shape and constrain their lives. Infrastructural work in this sense is both forward and backward facing, building and maintaining, and innovating and reproducing.

Infrastructuring after a disaster importantly rests on pre-event practices. In each of the historical moments in this book, I show how, even in destructive postdisaster situations, information infrastructures lend themselves to narratives of continuity: dominant institutions and information-related practices have momentum. When physical information infrastructure breaks, existing practices influence how people improvise and create interesting work-arounds as they seek to reproduce the information infrastructure that existed prior to the disaster.[22] Information practices are not for all time, and are forcefully shaped by particular institutional and technical configurations, but can be slow to change in that moment. Hence, as this book demonstrates, even though a disaster might create new information needs, the public information infrastructure sometimes remains relatively stable.

Ingrained in most thinking about information infrastructure is that, as an enabling resource, it is defined by reach beyond a single location.[23] Observers of early versions of the postal system admired the ways in which these wireless infrastructures transcended local areas, bringing news and opportunities from afar, much as pundits admired the early Internet's potential reach.[24] Information infrastructures produce and circulate objects called data, information, and documents beyond a specific geographic location.[25] People imagine that the reach of infrastructure means that all information will have the same effect in all places, whether it is in different geographic places or different social groups.[26] This line of thinking can lead to the conclusion that the reach of infrastructure is the only quality of infrastructure that is important, yet decades of research has shown that it is not just *access* to objects but also *local relations* that make meaning from objects.[27] So a picture of a cracked sidewalk can be circulated globally, but the meaning of this document will be different depending on the context in which it is viewed.

Yet the circulation enabled by infrastructure is still crucial to understanding the production of documents. The advent of the telegraph meant that people could follow the devastation of an earthquake from faraway places, and this ability to witness suffering as it happened changed people's relationship to aid because the urgency to act was so sudden.[28] An earthquake might be experienced locally, but the production and circulation of information beyond that particular location inevitably shapes the production of information about the event.

Public Information Infrastructures

In my formulation of *public information infrastructures*, the material entanglement of publics with information infrastructures is key to understanding their formation.[29] Different types of publics are shaped by particular material forms. Discourse publics can be formed around texts in circulation.[30]

Yet, it is not simply the text that is key to the development of publics—the manner in which texts circulate is also important. "Geeks" build "recursive publics" on the Internet to reflect their imagined "means of association," and Facebook, through its assemblages of data and algorithms, forms "calculated publics."[31] More broadly, material practices can constitute legitimate publics.[32] This book argues that information infrastructures can bring publics into being, and seeks to understand how information infrastructures shape the formation of publics and how publics affect the arrangement of information infrastructures.[33]

While there are many variations on normative ideas about publics, they are generally political formations, and I am interested in how publics are organized not just around information infrastructures, but also around events. From US philosopher John Dewey's perspective, a *public* is constituted around a specific issue—in this book, an earthquake. In Dewey's language, if the political issue is an earthquake, the public is involved in making sense of the shared associational consequences of the disaster.[34] People construct the world politically, and this also shapes how they experience a disaster. Following this, in my work, I examine earthquakes in California from the perspective of *issue publics* in order to highlight their political dimensions. Though the problem of making sense of an earthquake is not overtly a political process, it is in fact replete with power struggles.[35] In this sense, an issue public formed around an earthquake seeks to understand the shared consequences of an earthquake as a result of the decisions related to the distribution of suffering. It is also a political process to understand the damage done as well as the circulation of stories about the damage. Depending on the financial implications of a particular insurance policy, humanitarian aid effort, or business investment scheme, people might want to emphasize or hide the damage done by an earthquake—and these efforts are often partially accomplished through the production and circulation of documents. *Earthquake issue publics,* or more simply, *earthquake publics,* are constituted through the experience of an earthquake and are part of producing public information infrastructures. Earthquake publics are not passive entities; this book is concerned with how earthquake publics actively make sense of and direct the consequences of disaster through public information infrastructures.

Of course, the public is not a monolith, and neither are earthquake publics. The most vulnerable members of society who have been structurally marginalized are frequently the people who experience the most negative effects of an earthquake and are also usually without a voice in the earthquake publics that control how an earthquake is documented. In the case of the Chinese San Franciscans in 1906 or Spanish speakers in Watsonville, California, in 1989, non-English speakers used alternative approaches to getting aid and finding loved ones after a major earthquake. There was not a single information infrastructure underpinning a single public but rather alternative public infrastructures supporting multiple earthquake publics. Thus, this book argues that multiple sociotechnical assemblages—*alternative public information infrastructures*—constitute and are constituted through different earthquake publics, which are not dominant in the information order. As such, not everyone who was affected by an earthquake was automatically part of a singular earthquake public, and the earthquake publics constituted through public information infrastructures were not inclusive. To the extent that public information infrastructures were the domain where the shared consequences of the earthquake were documented, the ways people were and were not represented were crucial to their experiences of the earthquake. The multiplicity of alternative public information infrastructures and its overlaps points to the politics of access and exclusion.

Differential Vulnerability

In disaster research, the "vulnerability" approach pioneered by geographers and anthropologists stresses that different groups of people often have different experiences of a disaster.[36] Research in the vulnerability tradition has reiterated that the communities that suffer most in disasters are the vulnerable and those whose interests are marginalized and ignored. Sociologist Eric Klinenberg refers to his heat wave analysis as a "social autopsy," calling attention to the fact that although heat-related complications were the official causes of death, social conditions "made it possible for so many Chicago residents to die in the summer of 1995."[37] While it might be tempting to simplify "social conditions" or the descriptors of vulnerability, these communities must be understood in relation to the forces of capitalism and as involved in complex "double binds."[38] The lens of "differential vulnerability" has helped disaster researchers and policy makers understand variations in the impacts of disaster. My interest in differential vulnerability is primarily as it relates to public information infrastructures.

Similar to disasters, differential experiences of infrastructure are bound to status and privilege.[39] Geographers Simon Marvin and Stephen Graham's *Splintering Urbanism* delves into uneven experiences of infrastructure: they describe infrastructural holes, the ignored places with little or no infrastructure, and the ways in which infrastructure is classed, such that some have abundant access to high-quality infrastructure while poor and marginalized people must live with infrastructure that barely functions.[40] Marvin and

Graham show that infrastructure in the United States is increasingly owned by private infrastructure firms focused on profits, not universal access or quality. These companies frequently bypass poor areas unless required by law, thereby creating infrastructural enclaves for the wealthy.

Researchers focused on information infrastructure have noted differential experiences of infrastructure in examining everything from how the types of categories embedded in an information standard enforce heteronormative gender rules to Susan Leigh Star's idea of infrastructural "orphans"—the people left out of an information infrastructure.[41] For example, the very poor, especially those who are transient, may be unintentionally orphaned from public information infrastructures that many would assume to be inclusive, because these infrastructures require an address that can fix people in space in order to participate. Public information infrastructures are necessarily going to have limitations that affect how inclusive they are—sometimes this is by design, and at other times, it is less intentional.

Similar social, political, and economic forces underpin both differential vulnerability and public information infrastructures. As the eminent disaster sociologist Kathleen Tierney implores, "I must emphasize again that vulnerability to hazards and disaster is inseparable from more general patterns of oppression and marginalization that play out on a daily basis during nondisaster times."[42] New research looking at the intersection of social media platforms and disasters has shown that the distortions and inequities present in "digital divides" among technology "haves" and "have-nots" further discriminate against low-resource peoples after a disaster.[43] This reinforces the idea of postdisaster differential vulnerability as well as the theses of *Splintering Urbanism* and infrastructural orphans. Thus, the question of how information infrastructures form the membership of earthquake publics must be front and center.[44] By attending to the differential consequences of infrastructures and crises, this book underscores, in a historical context, that the sociomaterial relations that underpin postearthquake public information infrastructures are intimately bound to what kinds of earthquake publics emerge in response to earthquakes. Using the conceptual language outlined in this introduction: the information order, through the articulation of multiple public information infrastructures, enables and shapes event epistemology and, crucially, the kinds of earthquake publics that emerge in response to disasters and the activities that earthquake publics undertake.

For the cases of the earthquakes explored in this book, the next section explores the research opportunities and challenges that result from the limitations of public information infrastructures. As I discuss below, I assume that differential vulnerability is going to be baked into the archives of an event, reflecting the evolving preservation goals of a variety of memory institutions in different eras.

Research Approach

The importance of becoming informed after an event such as an earthquake is perhaps obvious or oversimplified, but it is repeated again and again in the disaster literature. I also find this a useful starting point because it helps to elucidate why disasters might be a particularly compelling site to investigate information orders. Before I dive into the question of why study information orders and disasters, I want to address why one might study information and communication practices in disaster situations at all—they can seem trivial compared to loss of life and other forms of destruction. Indeed, while documents are central to this study, I don't wish to suggest that it is the most significant thing about an earthquake event; from the perspective of reducing fatalities, the built environment is key. Information-related activities, though, are still crucial to disaster survivors and those who care for them. Researchers show that after a disaster, people first secure their safety and that of those around them, and then try to find out what has happened to others.[45] Making sense of what had happened, in the eras and places of the disasters explored here, were information-intensive activities. Understanding the information orders can illuminate what earthquake publics knew.

Why study information orders after disasters? First, disasters are an occasion when the infrastructures for producing and circulating documents can gain visibility because people are eager for information about what has happened generally along with the fates of loved ones. There is not just an intense need for public information but also the possibility that things are not working as expected. People reveal how they believe infrastructure *should* work when public information infrastructures are used intensely, overwhelmed, or broken. Disasters, as a time of irregularity, offer a window to interrogate public information infrastructure.[46] The circulation of documents, in addition to the documents themselves, shapes what people know about a disaster. And in some sense, this is why it is so fascinating to study postdisaster information orders. Because the silent structuring work of infrastructures—so integral to modernity—becomes easier to "see" after disasters.

Second, studying disasters gives us a window into the durability of modernity. As this introduction discusses, public information infrastructures both embody politics and produce a kind of politics. Some scholars take this a step further: as complex sociotechnical assemblages, infrastructures are ideological vessels that confer meaning on the societies that produce them; functioning infrastructures are symbols that a society is modern and progressive.[47] Infrastructures are not only symbolic of modernity; they are crucial for ordering and stabilizing it.[48] When public information infrastructures break, it occasions not only a dearth of information about what has happened but also a public's angst about its status as a modern society. As this book will illustrate, public information infrastructure has an especially intense symbolic value after an earthquake. Its breakage is not merely an inconvenience; broken public information infrastructure is a source for interpreting a disaster and the modern character of a society. And because of infrastructure's importance to maintaining modernity, new forms of governance have arisen to protect infrastructure.[49] Throughout the book, I examine these institutional arrangements in the space of public information infrastructures.

Taking this a step further, Kevin Rozario contends that destructive events, followed by renewal, are integral to modernity. He calls these phenomena the "catastrophic logic of modernity," and uses that phrase to explain our fascination with the spectacle of disaster and subsequent suffering.[50] The catastrophic logic of modernity is rooted in the economic concept of "creative destruction." Traditionally, this notion hypothesizes that crises are opportunities for the markets to work more efficiently; newer technology replaces old technology and leads to growth.[51] In popular discourse, extreme versions of the creative destruction hypothesis applied to disasters invoke an argument that posits that disasters rid society of weakness—for a Darwin-like survival of the fittest. And indeed contemporary social theorists assert that disasters provide opportunities for the private sector to further enhance its agenda through "disaster capitalism."[52] While concerns often revolve around the expansion of state emergency authorities in crisis moments, throughout the book I note that similar scrutiny is needed for private companies in the information and communication sector outside of postdisaster moments. Given infrastructure's role in producing modernity, events that threaten infrastructures are particularly salient because they reveal the arrangement of powerful institutions that can survive a disaster.

Locating Information Infrastructure

The preface of this book looked at the information order of 2011 Japan and how it shaped people's experiences of the earthquake. It is an example of the specificity of an event epistemology and its consequences. When I started this book project, I was interested in disasters that shaped my community at that time: earthquakes in the Bay Area. The four moments that this book examines all occur in Northern California, a place strongly identified with earthquakes, though I could tell the story of postearthquake information and communication practices elsewhere in the United States, or even the world. I chose to anchor this story to a specific region because infrastructures are made meaningful in places. The fact that the eyes of the world relied on the functionality of the postearthquake infrastructure was not lost on the people of the Bay Area during these earthquakes. Moreover, infrastructures can also facilitate objects crossing space in ways that raise important questions about the meaning of place. Yet even as infrastructure makes it easy for objects to move through or even transcend place, place will always remain a critical site of fixed capital and resources. I seek to locate infrastructure in the Bay Area during and after the earthquakes, while also paying attention to the ways in which infrastructure is global.

The informational flows that infrastructure facilitates uniquely shaped the production of local information after these earthquakes. The "fixity" of information, as it is locally produced and circulated, matters.[53] The specificities of the geography of the Bay Area—with the major city of San Francisco surrounded by water on three sides, wealth accrued in cities as a result of the extractive industries, racialized "frontier" ideology, policies of San Francisco and the state of California oriented toward economic development, and of course, dominance of information and communication industry corporations—all contributed to the particularities of the different postearthquake information orders. It is a region where many industrialists profited from the information and communication infrastructure: first as newspaper owners, such as the de Youngs and Hearsts, and later as leaders in the high-tech industry with names like Hewlett, Packard, and Zuckerberg—topics of many books unto themselves. The Bay Area is a site of earthquakes as well as a site of the production of the information and communication infrastructures that cross the country and world. This makes it an especially fascinating place to locate research that combines the two.

Documents as/about Histories of Information Infrastructure
In this book, my objects of study are the documents produced about the Northern California earthquakes of 1868, 1906, and 1989, and contemporary earthquake disaster response planning documents. My scope includes media stories, letters, pictures, government documents, scientific reports, and research papers. As I discovered, read, and learned from the documents, I attempted to consciously construct an archive as a field site.[54] Studying

information orders in historical moments is also an exploration of various archives, themselves sense-making resources. My work on public information infrastructures is embedded in institutions of memory, and I am limited by the availability of archival materials. To some degree all histories have a history of an information order embedded within them; in the Western documentary context, elites often shape what gets documented and what documents are preserved.

The documents that I consult are often about breakdowns in the ability to circulate documents. Examining infrastructure at times of breakdown is one well-known way of "seeing" infrastructure.[55] Dianne Vaughan's study of "organizational deviance" uses historical documentation to understand the many contingencies involved in system breakdown.[56] In my study of documents related to earthquakes, I borrow from Vaughan's approach to "historical ethnography," particularly bringing an ethnographer's sensibility to considerations of culture and what goes unrecorded.[57] Studying infrastructure in disaster moments—when infrastructure is broken and reassembled—is a kind of elaboration of this approach.

In addition to examining infrastructures at the time of breakdown, infrastructure researchers also invert the infrastructure to "look at the 'bottom.'" Paul Edwards argues that climate scientists are always inverting infrastructure to understand where climate data come from—as a process of "continual self-interrogation, examining and reexamining its own past."[58] After a disaster, when infrastructures do not function in the ways that people expect them to, people frequently participate in some type of infrastructural inversion as they consider how to make public information infrastructure work as they wish it would.

Star suggests another way of seeing information infrastructure. For Star, studying information infrastructure means reading a document as an artifact, record, and veridical representation of infrastructure.[59] Each of these different methods implies a different way of studying information practices in a historical context. For example, newspapers as *artifacts* can be read in various libraries and archives, and sometimes have a material quality to them different than the original artifact. The newspapers have articles that can be read as *records* or *traces* of what people thought or what happened. But I can also understand newspaper articles themselves as *veridical representations* of people making sense of the earthquake. In this mode of reading, the multiple materialities of the various veridical accounts of disaster, the documents themselves, in some sense are the history of information; they are an information order. (Or at least the documents are the available representation of the information order that I have to work with today.)

Documents help researchers see the information infrastructure; the documents themselves can be understood to be artifacts of the information infrastructure; documents also describe how the information infrastructure worked. For example, I use letters from people who have experienced earthquakes to understand earthquake publics. The letters provide accounts that are sometimes in contrast to the accounts of earthquakes found in newspapers or other professional reports, but sometimes these letters appear in newspapers. The fissures, references, and repetitions among the documents show how different accounts come to dominate, and who was constituted in various publics (and also make it more difficult to find what was discounted or entirely undocumented). Documents produce event epistemologies that point toward the information orders they entail and describe.

I found the orderings of the various public information infrastructures emerged as I engaged the documents and archives. Throughout this book I explore earthquakes in Northern California, but the compilation of these stories of these different earthquakes inevitably bears my own hand. My descriptions reveal a theory of what comprises the public information infrastructure and how this shapes event epistemology.

Four Moments

The structure of this book invites comparative conversations about historical event epistemologies in order to help us further understand how information orders constrain and service what we know.

I begin with the 1868 Hayward Fault earthquake, which occurred on the Hayward Fault that runs through Berkeley. This chapter destabilizes many of the contemporary assumptions about how disaster response is supposed to work: no government-led disaster response apparatus existed, some people weren't even sure if earthquakes occurred regularly in California, and the scientific community had no agreed-on explanations of earthquakes. After the 1868 earthquake, the chamber of commerce sent telegrams to newspapers around the United States and Europe, seeking to reassure outside investors that the earthquake damage was minimal. The stories of earthquake damage published in faraway newspapers, however, indicate that despite the chamber of commerce's efforts to use the affordances of the public information infrastructure to limit stories of damage, not all accepted their authority. Closer to home, local newspapers came up with their own sensational narratives about the earthquake and published the theories of locals explaining what happened. Ranging from weather that dictated earthquakes to gases escaping from the earth, local newspapers printed different earthquake theories for weeks after the earthquake.

Next, I jump forward almost forty years, and examine the famed 1906 earthquake and fire. The quake and subsequent fire destroyed a significant portion of the city, and forced thousands of people and businesses to relocate. In this chapter, to examine the information order, I concentrate on how family members, loved ones, and institutions accounted for people. I analyze how people used the telegraph and postal infrastructure to get in touch with others, and how registration bureaus attempted to organize the process of locating people. Public information infrastructure failures often led to creative improvisations. The post office initially relied on preearthquake sorting techniques, but gradually adopted innovative postmarking and sorting techniques for delivering mail to accommodate the unusual circumstances. The powerful mainstream press was the central broadcasting location for personal whereabouts. Work practices, powerful institutions such as newspapers, and early twentieth-century progressive ideas about data collection guided the postdisaster information practices. Nevertheless, the dominant public information infrastructures often excluded the poor and non-English-speaking residents of the city.

Out of the Cold War came organizations tasked with planning for and aiding in disaster response, such as FEMA and California's Office of Emergency Services. When the 1989 Loma Prieta earthquake struck approximately sixty miles south of San Francisco, these agencies sprang into action. In chapter 4, I focus on the government institutions responsible for planning around and responding to the earthquake. The federal, state, and municipal disaster response plans described scenarios where officials would collect information about the earthquake from other government workers, and then promptly disseminate public information via the media. In reality, the government itself learned about aspects of the earthquake from television and radio. Yet in a circular fashion, media reports frequently cited members of the government as sources. After the earthquake, the disaster response apparatus and media had a symbiotic relationship, each reinforcing the other's centrality. Many Bay Area residents questioned this centrality, though, criticizing the national media for alarmist reporting of the earthquake. I argue that the government's postearthquake information order reflected its vision of California as English speaking and middle class; in reality, poor and non-English-speaking Californians were the most affected by the earthquake.

The last chapter presents our current moment, the information order of today, and the public information infrastructures that anticipate responding to an earthquake in the Bay Area.[60] I examine disaster response plans from the city of San Francisco, state of California, and numerous US federal agencies to understand how the government envisions producing and making use of information for the public. In addition to the disaster response plans, I analyze information and communication technologies that companies intend for citizens to use postcrisis. In synthesizing these strands of the information order, I contrast government disaster response plans, which imagine producing authoritative information from the singular voice of the government, with the ways in which imagined communication practices are inscribed in the design of information and communication technologies that allow for widespread participation as well as imagine distributed authority. While both these entities inscribe a particular earthquake public in their plans, a dialectical relationship exists between the information production and circulation practices of the government and those of social media platforms.

I conclude by considering the historical moments in comparison and answer some of the questions driving this project: How do the information orders in different moments shape event epistemologies? What has changed, and what has remained unchanged or stayed the same? Within each historical moment, how did the information order change after the crisis? While these moments individually paint a conservative picture of infrastructural change—even the most creative work-around in moments of upheaval is often an attempt at trying to reproduce contemporary information orders; observing the differences between the information orders in these four moments makes it clear that great change has occurred. While a vast policy assemblage importantly shapes today's postdisaster information order, the telegraphic infrastructure critically influenced the event epistemology of 1868. That is, the constitution of the information order is always in formation: aspects of the order wax and wane in significance. I particularly focus on what I call the *bureaucratization of disaster response*— the creation of vast sociomaterial infrastructure that anticipate disasters. And yet as I look both forward and back, I find that many of the information practices are familiar to us, as are the power dynamics at play among the different actors producing the information order.

2 The Production and Circulation of Earthquake Knowledge in 1868 California

William Henry Knight wrote to his mother four days after the great earthquake of 1868: "You will have heard all about our great earthquake, the exagerated [*sic*] reports, and the succeeding reports making light of the whole affair. But a few words about it direct from one who experienced it may have a peculiar interest." The next three pages of his letter provide details about the earthquake as Knight tried to make sense of what had happened in San Francisco:

> Last Wednesday morning at five minutes to 8 ... the house ... was shaken as by a giant. ... We all suddenly adjourned to the street—not because we were scared, of course, but we wanted to see if our neighbors were alarmed. ... I ... immediately hurried to the store. The business streets were full of excited people, and rumors of killed, and fallen walls, etc. were rife.[1]

Knight wrote to his mother in New York to assure her that he was well amid conflicting stories already sent on the cross-continental telegraph and published in the newspapers; he was anxious about the tales that might have preceded his own. People such as Knight knew that when they wrote to their mothers to tell their versions of what had happened, they also needed to correct both the understated and sensational stories about the earthquake already in circulation.

The Hayward Fault ruptured on October 21, 1868, at approximately 8:00 a.m. Different reports today say thirty people were injured or died as a result of the earthquake.[2] The earthquake had an epicenter in Alameda County, near the towns of Hayward and San Leandro, and had an estimated magnitude of 6.8 to 7.0.[3] Although the earthquake was in the East Bay, many of the damaged buildings were in San Francisco, particularly on artificially filled land. The population of California had grown significantly since 1857, but Alameda County was still sparsely populated. The 1860 census listed the population at 8,927, and in 1870 the population was 24,237.

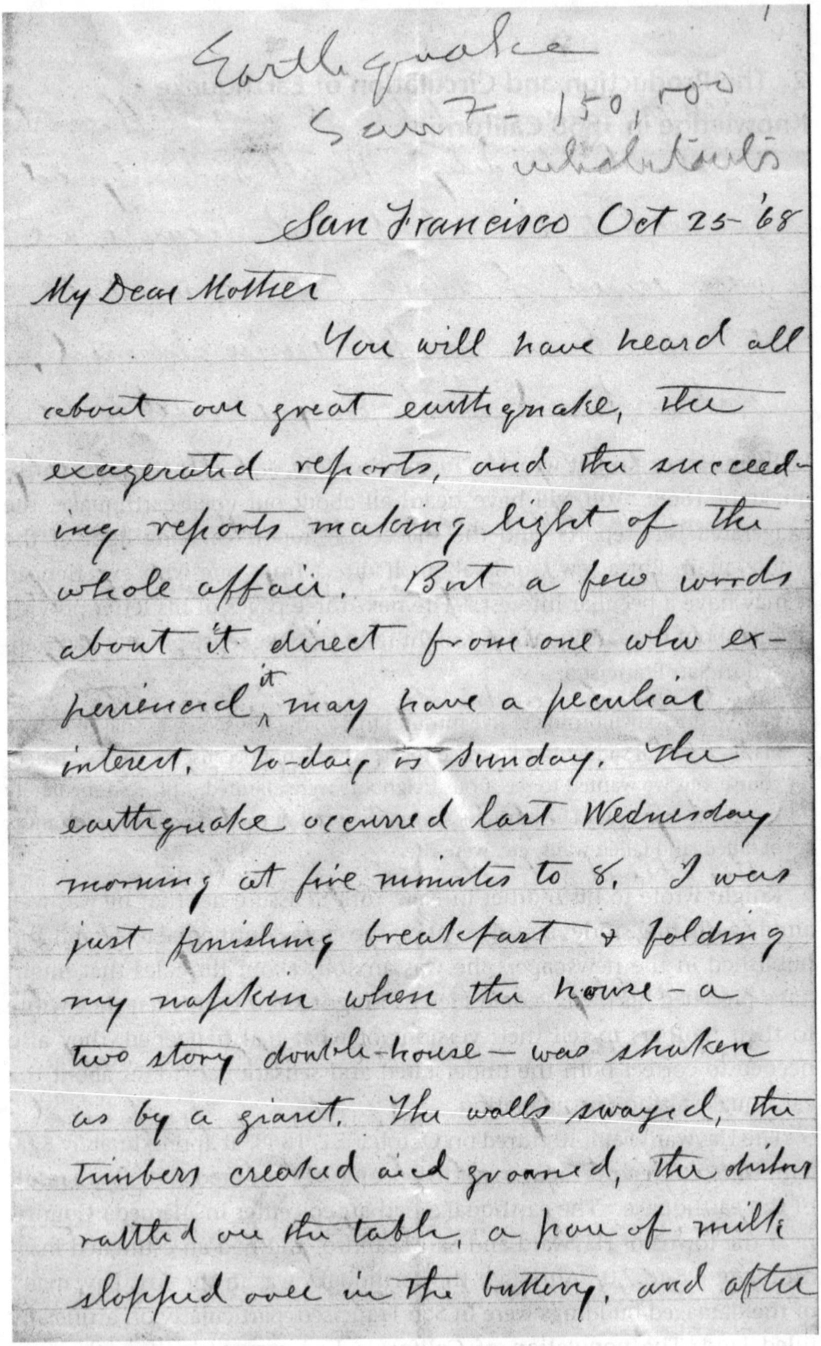

Figure 2.1
William Henry Knight to his mother, October 25, 1868. Image: William Henry Knight Papers, BANC MSS 761116 c: letter: 1, page 1. Courtesy of the Bancroft Library, University of California at Berkeley.

Conversely, San Francisco County was more densely populated and grew from 56,802 in 1860 to 149,473 in 1870. The earthquake was also felt on the San Francisco Peninsula in San Mateo County, damaging the courthouse in Redwood City. At the southern end of the San Francisco Bay, the spire of the Presbyterian Church in San Jose fell. Despite this damage throughout the region, after the earthquake most Californians turned to San Francisco, the business and population center of the state, to see how the residents of the city fared. Approximately fifty buildings in San Francisco were "wrecked" or "badly shattered."[4] The earthquake destroyed many architectural details, such as cornices and other decorative embellishments. The damage done to San Francisco, in terms of 1868 dollars, was a subject of great debate at the time; estimates ranged between $300,000 and $5,000,000.

This chapter focuses on how the information order shaped how some people made sense of that earthquake. This information order included newspaper companies, printers, letters, telegrams, telegraph lines (working and inoperable), telegraph companies, photographers, and sketch artists as well as institutions involved in documenting and interpreting the earthquake, such as the municipal government and California Academy of Sciences.[5] The English-speaking, white, and male account of earthquake publics in this chapter primarily represents the voices of those who wrote for and owned newspapers, and only those newspapers that have been widely preserved and remain accessible today. California has been a multiethnic, multilingual, multireligion, and multiracial community, however, since well before it became a US state.[6] By carefully tracing how knowledge about the earthquake was produced and shared, I examine how the print and telegraphic infrastructure as well as other organs of public information practices, shaped certain, but not all, public narratives of what happened.

This story is indelibly embedded in the context of a population boom in the region of Northern California in 1868. From the vantage point of the relatively new state of California and expanding public information infrastructures, this chapter argues that the characteristics of the public information infrastructure, and its ability to circulate documents nationally, shaped the production of knowledge. People produced reports about what happened with faraway audiences in mind. I explore the different ways in which understandings about the earthquake were produced—by reporters roaming the streets, through exchanges of telegrams reporting about the damage in a town, via announcements about fire safety, and in scientific reports about the earthquake—to show the heterogeneity of infrastructures that produced knowledge about the earthquake. Through these practices and the public's own interest in learning about the aftermath of the earthquake, the earthquake public and public information infrastructure were mutually constituted. Various entities—earthquake publics, business owners, newspaper companies, and the government—all had their information agendas and attempted to produce information that met their own needs.

The reader will note that institutions we might expect to be involved in defining what happened—seismologists and professional disaster responders employed by the state—are not part of the earthquake public's narrative of disaster because these categories of earthquake expert did not yet exist in California.[7] This chapter destabilizes assumptions that one might make about how disaster response was "supposed" to work: no government disaster response apparatus existed, and there were no agreed-on scientific explanations for earthquakes. In light of these unavoidably presentist observations, I look at an information order that produced the contemporary event epistemology through a close examination of the postearthquake information practices of English-speaking Californians. Residents of California were aware of earthquakes, and many recognized them as an inevitable consequence of living in their particular location, yet at the same time they considered them frightening and, among other things, bad for business. Citizens confronted earthquakes not solely from the perspective of their local experience but also from the perspective of the documents about the earthquake that circulated nationally, such as newspaper articles or the letter to Knight's mother.

By focusing on postdisaster information production and practices, this chapter builds on research by California historian Charles Wollenberg and architectural historian Stephen Tobriner. According to Wollenberg, elites in San Francisco wanted to downplay earthquakes to protect real estate value, and ensure that California remained a destination for capital investment and labor: "San Francisco businessmen took action to understate the extent of the damage and to assure both the local populace and outside investors that earthquake destruction was preventable. As part of this effort, the city's economic leadership sponsored an attempt to study the quake and find ways to prevent massive damage in the future."[8] Focusing on the efforts of some business elites overlooks what impact the earthquake might have had on local practice.[9] As Tobriner shows, though, architects and builders in California actually learned a great deal from the 1868 earthquake, and their subsequent building practices reflected new understandings of buildings during earthquakes.[10] Moreover, resistance to stories minimizing damage came not only from Californians interested in understanding the damage in order to learn how to build better structures but also from those who could profit by telling a different story. In exploring this, I build on Wollenberg and Tobriner's arguments by concentrating on the production and circulation of narratives about the earthquake.

The account of public information infrastructure throughout this chapter shows how an infrastructural account might overlap with an economic account—commercial interests are paramount to explaining the different stories about the earthquake that circulated—and political account—policies and political conflict in as well as among cities shape the stories—but these accounts don't capture how the sociomaterial qualities involved in the production of public information infrastructure shape the experience of the earthquake. This chapter follows the news of the earthquake from the earthquake-affected cities and towns to the wider population of the United States. I then examine the reporting and documentation practices of government and scientific organizations. These institutions mostly failed to produce the authoritative documents that the earthquake public wanted in order to help understand the earthquake. To borrow the terminology of Olga Kuchinskaya, after the earthquake the information order produced invisibilities as well as visibilities.[11]

Reporting on Local Damage

The earthquake public desperately wanted reports of what had happened after the earthquake, and the ongoing processes involved in producing and circulating documents, both the ones that worked and the ones that didn't, shaped the public's understandings of what happened. The earthquake public's desire for news about the aftermath was manifested in and through the public information infrastructure. Newspapers were critical for meeting earthquake publics' interest in learning about the impact outside their area. There was significant demand within San Francisco and surrounding towns for news about the earthquake as soon as the shaking stopped, but newspapers had to first contend with shaken type cases and typesetters before they could restart the presses. Within the cities, the infrastructure for producing information immediately after the earthquake consisted of reporters roaming the streets of the most damaged areas and cataloging their observations, editors assembling that work, and the print staff and printing presses owned by publishers then printing the news.

The *San Francisco Evening Bulletin* kept their presses running late, "so great was the demand for information in regard to what had happened."[12] To fulfill this demand, newspapers reported on the damages in their own cities and other affected areas along with the well-being of other people. In

San Francisco, where most of the damage occurred, newspapers advertised that their reporters had collected all the information regarding the damages to various buildings and people who might have been hurt. Generally, each of the city newspapers reported the local damage in great detail, describing each specific damaged building, business, and person. Sometimes the descriptions seem to be organized by location, as if personnel from the *Daily Morning Chronicle* had walked the streets noting all the damage they saw or heard of, and then printed the notes directly into the newspaper. Other newspapers used similar protocols. The *San Francisco Morning Call* explained, "The lengthy report of the calamitous event ... has been collected by faithful and reliable reporters, who speak from personal observation. It will be found nearly correct in detail."[13]

On the evening of the earthquake, the new *Daily Morning Chronicle* (which was the *Dramatic Chronicle* until early 1868) printed an *Extra* issue, and the editors were quick to pat their own back with the announcement that this was an act of "unparalleled journalism."[14] The *Chronicle* bragged, "Our account was so full [in the *Extra*] that the evening papers fell far short of it in completeness of detail." The *Extra* included long lists of incidents such as "a fire wall on Jones alley fell, but injured nobody," and "a row of two-story brick dwelling-houses on the north side of Howard and Third was considerably injured." Each incident was gathered by the *Chronicle* staff and captured in a few sentences.

Printing a newspaper after the earthquake was difficult. In some newspaper rooms, type was strewn all over the floor. As the earthquake took place just after 8:00 a.m., morning newspapers had already been published, giving evening newspapers the scoop. The evening papers, however, had to produce an edition with shaken printing rooms.[15] Traumatized typesetters and others who worked in the print rooms were reluctant to return to work. Yet because of the desire for extra editions of the newspaper, there was additional pressure for workers to return to the office.[16] Some damaged offices, such as that of the *San Francisco Evening Bulletin*, managed to print an evening newspaper the day of the earthquake.[17] Not all were like the *Bulletin*, though; newspaper companies that were in badly damaged towns or San Francisco companies that printed non-English newspapers recovered more slowly, likely due to being smaller and having fewer employees.[18]

Not surprisingly, given the origination of the earthquake, some reports said the most damaged newspaper was in San Leandro: the *Alameda County Gazette*.[19] The newspaper did print an edition several days after the earthquake, declaring, "We ... make no apology for our somewhat demoralized appearance, for we are thankful as we look upon the wreck around us, that we are able to issue even these few hastily written words."[20] Newspapers were proud of getting their publications out on time, so publicity about the disruption of printing was contentious; newspapers often offered sympathy to damaged brethren.[21] In an atmosphere where news was in great demand, among the newspaper companies working in damaged areas, the large San Francisco papers seemed to either suffer the least damage or recovered the quickest. News about damage in Alameda was circulated first by the San Francisco dailies and then by the local weeklies once they had recovered. The practice of producing records about what had happened was undertaken in a chaotic atmosphere, yet many newspapers were able to persevere, continuing to sell papers and fulfilling the earthquake public's desire to learn about what had happened. Still, the production of public information infrastructure was greatly affected by the newspapers' ability to recover—which was uneven depending on the resources of the newspaper and proximity to the earthquake epicenter.

Newspaper reporting about earthquake-related deaths was confusing, and reflected the chaotic and challenging process of collecting news in tragic circumstances. As an example, the *Daily Alta California*, an old and respected newspaper of the California establishment, reported on the number of deaths on October 22 that "four persons were killed by the falling of cornices and chimneys," but the newspaper seemed to include reports of five deaths.[22] Within three days, newspapers agreed on the number of dead as four in the city of San Francisco and one in San Leandro.[23]

Reports about physical damage to buildings were interspersed with news about injured people and even the dead as well as social commentary assessing the emotional state of the earthquake survivors. The mass of people was described as, at once, terribly frightened and calm; laughing without a care in the world and running into the streets, terrified; panicked, brave, and noble.[24] Newspapers even went as far as encouraging boastful nihilism in the face of earthquakes: "Californians are remarkable for their disregard of human life."[25] Social practices of 1868 shaped the manner in which people's reactions to the earthquake were described. Newspaper reports of gender-based behavior tell of emotional women as impediments to the reasonable reaction of men.[26] In some cases, women were apparently so frightened that they could not stay conscious; many articles reported women having trouble staying on their feet.[27] Not only were women portrayed as useless after the earthquake, but the fact that the earthquake occurred early in the morning meant that women were not properly dressed for the day—an apparently titillating experience for many men writing about the earthquake.[28] A popular trope was that of women being *dishabille*, meaning undressed; the newspapers usually announced this with great pleasure and much winking that "ladies" were found in such a state.[29] The sight of discombobulated and disoriented women was often portrayed as quite funny and even ludicrous, and shows how even a dangerous situation such as an earthquake had to fit into the dominant gender narratives.[30]

The reports produced about the earthquake were also created in a competitive environment where newspaper companies aimed to entertain in order to sell both their newspapers and their own cities. When people in San Francisco worried about damage in Oakland, the *San Francisco Daily Morning Chronicle* sent reporters to Oakland. The *Chronicle* asserted the day after the earthquake that the impact was dramatic: "To add to the terror of the scene, large trees ... were uprooted."[31] These reports outraged the Oakland newspaper reporters and editors: "The eyes of the *Chronicle* reporter must have been remarkably sharp, for he saw what no other person in the city saw if he did see any of our oak trees uprooted from their beds."[32] The *Chronicle* argued, "The *Oakland News* is trying to make out that there was no earthquake in Oakland worth mentioning."[33] Oakland newspapers countered by saying, "[Visitors from San Francisco] had been reading the *Chronicle*, and expected to see things here in a ... worse state."[34] The debate between the *San Francisco Daily Morning Chronicle* and Oakland newspaper companies is emblematic of the character and humor of some of the newspapers of the era—not only of the *Chronicle*'s particularly "sensationalistic" style as an upstart paper for the working person, but also of the regional rivalries at stake in the reporting of the earthquake, the latter of which shaped how the earthquake was reported. No city wanted to appear as the most dangerous to live in, and all wanted to appear to be "safe" from future earthquakes.

Cities tried to bolster their image after the earthquake, often at the expense of others. As one example of this, the *Oakland News* said that San Franciscans were buying homes for themselves in Oakland as a safety precaution.[35] To combat the negative perception of residents buying property elsewhere, the *Alameda County Gazette* asserted that "people will not be frightened away by earthquake," and printed advertisements claiming that the "Real Estate office of G. E. Smith" was "crowded" with San Franciscans buying "farms and homestead in Alameda County."[36] San Jose, fifty miles to the south, also claimed that people were buying property there in favor of San Francisco.[37] News making about the earthquake took place in an environment of competition among newspapers and regions. The process of news making also reveals the ties between the press and local politics. The earthquake was far from being a moment when social, political, and economic norms were disposed of. As actors in the complex information order, male reporters pursued their own information agendas by imparting their observations of damages and recovery activities within the cultural practices of San Francisco in 1868.

Earthquake Damage Elsewhere

Newspaper editors and reporters witnessed as well as documented the physical damage in their cities, and perhaps nearby cities, on the day of the earthquake, but people were eager to ascertain the earthquake's impact around California. Where else had been affected? Was the damage worse elsewhere? Were people OK? Thus, newspaper companies made use of the telegraph to learn of the fates of nearby towns, which they would then broadcast in print. In regular circumstances, the large daily San Francisco newspaper companies relied on the telegraph for national and international news. After the earthquake, San Francisco newspapers turned to the telegraph to find out what was happening in other California locations that may have felt the earthquake, and people around the United States turned to newspaper and telegraph companies to learn about what had happened. Most of the local newspapers reported on damage in other cities based on information from telegrams, usually grouped by region or county. This made it possible to ascertain the reach of the earthquake and relative impact in different regions.

Newspapers promised that their telegraphic dispatches were unique, but nearly identical reports under the headline of "Telegraph" appeared in popular San Francisco daily newspapers including the *Daily Morning Call*, *Daily Morning Chronicle*, and *Daily Alta California*.[38] The first reports of faraway damage printed in newspapers were received by telegraph, but detailed reports published in subsequent days were received via letter or eyewitness report, or else copied from another town's newspaper. Most newspapers copied stories from other papers, including those that were collected via telegraph—sometimes wholesale, and at other times, borrowed in bits and pieces.[39]

Yet however similar these stories were, the newspapers, as public information infrastructure, were not a uniform entity. The *San Francisco Evening Bulletin* was the most probusiness.[40] On the other end, the *San Francisco Daily Morning Chronicle* was the most "sensational," having been started only a year earlier by the teenage de Young brothers.[41]

Furthermore, other differences in the newspapers came about because the telegraphic infrastructure did not provide flat access for telegrams for all newspapers. Almost a decade before the 1868 earthquake, the US Congress passed the Pacific Telegraph Act of 1860, and the cross-country telegraph was completed in 1861.[42] The telegraphic infrastructure that had connected parts of California was then also connected with the rest of the United States. The local California press celebrated the new infrastructural addition to California: "The lightning has annihilated a continent as an obstacle to *intellectual communication.*"[43] These declarations were not new or specific to California. Popular rhetoric imagined that the telegraphic infrastructure improvements would mean unconstrained public understanding and communication. In reality, the telegraph was expensive, monopolized, and fell short of the goal of facilitating a public sphere. By 1866, much of the cross-continental telegraphic network was owned by Western Union, which charged high prices to use the telegraph network, making it mostly useful to businesses such as daily newspapers, which thrived on national news.

The newspaper and telegraph companies, as actors producing the public information infrastructure, were soon embroiled in complex economic relationships. Once the cross-continental telegraph was completed in 1861, three newspapers in California controlled the news from the East for almost a decade.[44] In California, the *Sacramento Daily Union*, and in San Francisco, the *Daily Alta California* and the *Evening Bulletin*, shared news digests before the cross-country telegraph was even deployed, delivered from the edge of the telegraph network in Saint Louis to San Francisco via the Pony Express.[45] These newspapers became known as the California Associated Press and received dispatches from the Associated Press.[46] These three California newspapers enjoyed exclusive access to Associated Press dispatches, but these dispatches were sometimes "stolen" through creative means by other newspapers.[47] One of the de Young brothers, the infamous proprietors of the *San Francisco Daily Morning Chronicle*, apparently deciphered messages by the sound of the telegraph alone and then published the messages in his newspapers.

After the earthquake, many of the same telegrams appeared in different newspapers, regardless of their association with the California Associated Press or other regional associations.[48] The intertwined economic relationships among the newspaper and telegraph companies, established long before the earthquake, greatly shaped the earthquake public's event epistemology and production of information.

Telegraph Offices as Sites for News

The telegraph companies were not mere conduits in the public information infrastructure. The way the telegraphic infrastructure worked—and after the earthquake, temporarily didn't work—was different in each community and served to shape the story as it unfolded. The telegraph offices were important sites for the earthquake public to not only learn what happened via telegrams but also interpret the public information infrastructure based on whether or not it was working as usual. The processes involved in continually producing public information infrastructure changed after the earthquake because of damage as well as the earthquake public's insatiable need for information. Public information infrastructures produced and circulated information about the disaster, and were constitutive of the information-lacking earthquake publics.

After people outside San Francisco felt the earthquake, they rushed to telegraph offices to learn about their businesses and brethren in San Francisco. A Sacramento newspaper said, "The telegraph offices were besieged" with people wanting to know the effect of the earthquake, "citizens rightly judging that if the shock was so heavy in Sacramento, its effects at the Bay City must be most disastrous."[49] Many telegrams were reportedly sent to San Francisco. People outside San Francisco imagined that their impulse to get in touch with loved ones in San Francisco was equally returned. Reports said that people sent "innumerable messages," but hardly any answers were received because the people in San Francisco "were too excited, undoubtedly, to attend to telegraphing."[50] The volume of telegrams to San Francisco was so great, the *Daily Alta California* joked, that a telegram inquiring about the well-being of a "Mrs. Smith" was answered: "Mrs. Smith all right; in capital health and spirits; sends her love," despite the fact that the recipient "did not have the slightest idea *which* Mrs. Smith was meant."[51]

Outside the Bay Area, telegrams were posted on bulletin boards as a way of broadcasting what happened. According to a local newspaper from Marysville, a town in the northern Sacramento Valley, a telegram from San Francisco was "posted on bulletin-boards, which was read by thousands during the day, and in the afternoon we received many calls from citizens who were anxious for later and more specific reports."[52] As post offices in California's gold rush era had served as locations to make contact with people far away, the telegraph offices after the earthquake were sites to hear about the news of the earthquake and hopefully learn about the fates of associates.[53] The gathering of earthquake publics at bulletin boards shaped the event epistemologies.

The telegraph offices were also sites of rumors; news exaggerating the damage in San Francisco spread all over the state. Some of these rumors said that San Francisco was destroyed and sixty bodies were recovered.[54]

Newspapers often blamed (or cited) the telegraphers as the rumor origina-
tors.[55] The narrative about the impact of the earthquake in San Francisco as
it unfolded via telegraph and newspaper in Virginia City, Nevada, on the
east side of the Sierra Nevada mountain range, serves as an example of the
rumor unfolding in the town. The local newspaper reported, "At first the
story ran that at least one-half of the city of San Francisco had been swal-
lowed up, and Oakland and other towns almost demolished."[56] After this,
people looked to the telegraph office for the latest news. But they found
that they were not getting their news from San Francisco. All the news was
coming from an operator in Oakland whose office was a wreck and had "cut
the wires" outside town, meaning that there was no news coming directly
from San Francisco.[57] In Virginia City, the story about the Oakland operator
"excited rather than allayed the general anxiety." After a direct connection was finally made between Virginia City and San Francisco, an "Extra
Edition" of the newspaper was issued. Yet this did not settle the excitement about the story: "A number of private dispatches" were received, and
"the excitement was continually kept alive—the more news, the more the
eager people clamored for news."[58] The *Oakland Daily News* later disputed
the story of the Oakland operator.[59] The *Daily Alta California* lamented the
impact of the first telegrams sent, blaming a "mischievous person" and the
telegraphers for "exaggerating every notable occurrence."[60] The newspapers
used the telegraph operators as a foil for misinformation.

The telegraph offices were important locations for inquiring about the
fates of other communities. When people did go to the telegraph offices to
find out what had happened, however, they were sometimes greeted with
rumors. The working telegraph might have been a conduit for mischief, but
the broken telegraph was cause for panic, as the skeletal version of what
happened could be filled in by a person's imagination. Thus, a community
whose telegraphic connection with San Francisco was severely damaged
interpreted the earthquake damage via the information infrastructure. The
infrastructure itself was a source through which to understand and make
sense of the disaster.[61] The telegraph promised that people would have
more immediate access to events in faraway places, but the existence of the
telegraphic infrastructure did not mean that the assessment of the events
that were circulated had any correspondence to truth: newspaper companies were doing some of the work assessing the validity of the telegrams, as
were the earthquake publics gathered at the telegraph offices.

The circulation of news was facilitated by imbricated sociomaterial practices: newspapers printed telegrams, telegraph operators sent over the wires
what they read in newspapers, and people traveled on foot, by horse, and

via boat with newspapers, letters, and telegrams.[62] All these means of circulating news are critical to understanding how the public information infrastructure worked after the earthquake.

When some communities in the East Bay tried to contact San Francisco, they found that telegraph communication had been cut off.[63] Similarly, a San Francisco newspaper reported that the telegraph was "not operating" the morning of the earthquake.[64] Faced with a nonworking telegraph and desperate to hear from San Francisco, Oaklanders tried to work around broken parts of the telegraphic network by sending messages from Oakland to San Francisco via Sacramento:

> Many inquiries were made at the telegraph office, on Seventh street, to know if any dispatches had been received, descriptive of the effects of the earthquake there, but for some reason the telegraph communication was not complete with that city [San Francisco], and it was some time before any dispatches were received. Even at a late period of the day the telegraphic communication between Oakland and San Francisco was by way of Sacramento. About nine o'clock intelligence was received that several lives had been lost, and that the destruction of property was very great in our sister city.[65]

While people in Oakland waited for the telegraph line with San Francisco to be fixed, boats crossing the bay between the two cities brought newspapers from San Francisco, which both sensationalized the damage and soothed worried Oaklanders.[66]

Damage Estimates

Estimates of earthquake damage in dollars are controversial even when done with ample time and explicit processes. Most of the numerical damage estimates that circulated immediately after the earthquake were inevitably inaccurate. Yet damage numbers became a shorthand way for people to debate the earthquake's consequences. In this analysis, they serve as a way of analyzing what kinds of stories circulated and to whom. Examining how the California newspapers estimated the earthquake damage for eastern audiences gives us an idea of how the reach of public information infrastructure shaped local conversations. A damage estimate put forth by the chamber of commerce seemed to try to underestimate the damage to San Francisco for those outside California. Some San Francisco newspapers joined together in promoting the city as a safe place to live and invest, but these efforts to convince people in the eastern United States, or even just Sacramento, were not necessarily successful.[67] The chamber tried to be the informational authority, but the process of estimating damage was clearly problematic, as the *Sacramento Daily Union* cautioned that "the authority of leading citizens and leading journals" was under suspicion.[68]

As Knight's letter in the beginning of the chapter suggests, there was general anxiety about how the earthquake was reported in the eastern United States. The East Coast discovered the news of the earthquake via three telegrams, which appeared in various forms in its newspapers.[69] Immediately, these telegrams included numerical dollar estimates of the damage. Many newspapers received a series of four telegraphic dispatches on October 21, the day of the earthquake, and some printed these reports verbatim the next day.[70] The first dispatch after the earthquake included no specific estimate: "At the present writing, 9 a.m., no estimate of damage can be made, though it is considered comparatively small."[71] The second dispatch was evasive about aggregated damage estimates: "The damage will not exceed one million dollars." Many newspapers reported this specific figure—either near or over "one million dollars"—in the days after the earthquake.[72] For example, on October 22, 1868, the *New York Times* ran the headline "Nearly a Million Dollars Worth of Property Destroyed."[73] Despite the attempts to avoid it, aggregated damage estimates were sought and speculated on immediately after the earthquake.

The San Francisco Chamber of Commerce was concerned about the implications of these reports to the East. The *Daily Alta California* reported that getting the "correct" story out was discussed at an impromptu Board of Supervisors meeting held on the day of the earthquake.[74] The chamber of commerce was also concerned: the third dispatch to the eastern United States on October 21, 1868, concluded, "The [San Francisco] Chamber of Commerce held a meeting to-day and resolved to telegraph to the Chambers of Commerce in New-York, Philadelphia, Boston, Chicago, London, Paris and Hamburg the account of the disaster."[75] As Wollenberg details, the chamber of commerce was alarmed by the early telegrams that had been sent, and was intent on countering the details of the damage with a telegram that would minimize the damage reports and hence concerns about the business prospects for San Francisco. The full text of the telegram sent out by the chamber is as follows:

A severe shock of earthquake, experienced here at 7:50 A. M. Considerable alarm felt at time of occurrence. A good many buildings on made ground injured. Custom House and City Hall, both poorly constructed, badly injured, and some buildings in process of erection have fallen in. Some parapet walls falling have caused the loss of four lives. No damage to well-constructed buildings. Total loss on property will not exceed $300,000.[76]

This was far less than other local estimates made in the days after the earthquake, which said that the damage could cost several million dollars.[77]

Whatever the intent of the chamber of commerce, its telegraphed report did not appear in the eastern newspapers alongside the first mentions of the earthquake and did not have the desired effect of defining the news story. In the weeks after the earthquake, few newspapers in the East used the chamber of commerce estimates.[78] Many others, however, continued to report ranges that included the chamber's estimates, but also extended up to several million dollars in losses; these newspapers often favored the larger estimates, reasoning that "the latter figures [of two million] are probably nearest to the true loss, as quite a number of the buildings have been torn down, and will be reconstructed."[79] The *London Times* printed a letter in which the correspondent said that the news and damage reports from San Francisco contained in these "unreliable telegrams" should not be trusted, since they were from people "desirous of suppressing as much as possible the disastrous effect and great damage done to property."[80] Later, the *Chicago Tribune* printed a letter "From Our Special Correspondent" that called the estimates by the chamber "simple absurdity."[81] Weeks after the earthquake, the *New York Times* said, "The destruction of property cannot be overestimated at $1,250,000."[82] It is not a surprise that the chamber of commerce could not control the story of earthquake damage. The *Journal of the Telegraph* estimated, "On 'earthquake day' in San Francisco, two thousand telegraphic messages were sent East."[83] The story was bigger than the chamber's reach.

Moreover, the chamber of commerce damage estimate number was even mocked by California newspapers outside San Francisco. While prominent newspapers, particularly the probusiness *San Francisco Bulletin*, vigorously defended the chamber's estimates, in areas affected by the earthquake outside San Francisco, newspapers called the estimates "extravagant" as well as "an affront to rational observation and common sense."[84] The respected *Sacramento Daily Union* was highly critical of the estimates, cautioning, "Men of capital are not going to be deceived by misstatements or gulled into confidence ... by under-estimates of damages even on the authority of leading citizens and leading journals."[85] It acknowledged that while "San Francisco cannot seriously suffer in a depreciation or loss in property without making all the rest of the State a sharer in the calamity ... it is easy to make up a detailed statement with much semblance of truth and disguising the whole truth."[86] California newspapers argued about the damage figure for weeks, with most newspapers outside San Francisco contending that the chamber of commerce number understated the real amount. The

Sacramento Daily Union called out the *San Francisco Evening Bulletin*, whose estimate supported the chamber of commerce, for not seeing that "losers" were giving false statements, whereas the *Sacramento Daily Union* estimate of "several millions" was "editorially made" from telegrams from uninterested parties.[87] The *Union* concluded, "It would have inspired more confidence abroad" if there had not been "a studied purpose of concealment and prevarication."[88]

Following the earthquake, news dealers in San Francisco published a pamphlet to examine the veracity of the chamber's estimates.[89] Using "personal inspection by our reporter" along with "estimates of the builders and owners," going block by block, building by building, the writers determined that the loss in the city would "not exceed $500,000."[90] The pamphlet concluded that the "estimate of the Chamber of Commerce was perhaps within the mark; but was a reasonable and judicious judgment at the moment."[91] Sensationalist newspapers and unscrupulous capitalists account for some of the dramatic variation in damage assessment, but the measurement of what exactly is "caused" by an earthquake is difficult to quantify in dollars— an ambiguity that people took advantage of. Surprisingly, as Wollenberg notes, despite the amount of damage being wildly debated at the time, a number close to that of the chamber's estimation shows up in popular reports today.[92]

While the damage assessments by the chamber of commerce were as preposterous as any other specific estimates, they did not catch on despite the assertion by the earlier dispatches that the chamber's estimate was authoritative. An assessment of the efforts of the San Francisco business elites shows that their message downplaying damage did get traction, but was not widely circulated or believed. What is notable is that simple estimates of damage—from $300,000 to millions—were so widely cited. These damage estimates were able to quantify to people "how bad" the earthquake had been in simple terms. Faraway locales chose to publish estimates that conveyed many magnitudes of difference from the actual dollar amounts. Local papers were also doubtful of the chamber of commerce's estimate, but instead argued about the methodology involved in assessing damage as well as the political motivations of those producing and circulating these numbers. The earthquake publics constituted through public information infrastructures did not produce harmonious accounts of the disaster. The reach of the telegraphic infrastructure coupled with the strong relationship between some telegraph and newspapers companies made newspapers a primary vehicle for various parties to struggle over the master narratives of the earthquake.

For Eastern Friends

Although some newspapers may have intended to downplay the damage or stick to the estimate given by the chamber of commerce, other newspapers, photographers, and printers were anxious to capitalize on the images of the damage. These actors sought to produce documents about the earthquake specifically for distant audiences, and these documents did not soft-pedal the damage. San Francisco was a growing city, and the telegraph provided a way for people to immediately hear news, but other media supplied actual details about what happened. Letters such as the one Knight wrote to his mother were one way that people could learn about an earthquake from loved ones far away. Another way was for people to send along newspapers designed for faraway audiences. The *Daily Alta California* ran advertisements for an issue of *Golden Era* "suitable for mailing to Eastern friends" that "gives a correct idea of the effects of the shock."[93] Newspapers were frequently sent through the mail to personal acquaintances—a practice held over from before the Postal Acts of 1845 and 1851 (which lowered the cost of sending letters).[94] Even well into the 1860s, newspaper companies advocated that the newspapers were not just for broadcast but could also play a part in personal communication to friends in the eastern United States. "Steamer" versions of newspapers were often made by larger newspaper companies for the explicit purpose of summarizing news for another locale and published at weekly intervals. After the earthquake, many newspapers published special issues explicitly designed for sending outside California. The *Daily Alta California* created a smaller sized "Half-Sheet ... Steamer Alta" for nine cents that had a "full and complete account" of the great earthquake.[95]

People in the eastern United States also learned about the earthquake through imagery. With the few exceptions of "illustrated" newspapers, daily newspapers did not regularly have images. After the earthquake, however, printers in the Bay Area produced a variety of publications to send to "Eastern friends" that included images of the earthquake, often in the form of sketches printed in periodicals.[96] Letter sheets with illustrations of the earthquake damage circulated; an advertisement immediately after the earthquake advertised "Appleton's Letter Sheet," which had "illustrations and a complete account of the recent Earthquake," stating that the letter sheet was "the most convenient form to send East."[97]

The *Daily Alta California* also advertised that it "received from D.E. Appleton & Co., photographs of the Court House at San Leandro, and the warehouse and mill at Hayward's—which were destroyed by the earthquake."[98]

SAN LEANDRO COURT HOUSE, Alameda Co. as left by the Earthquake of Oct. 21, 1868.

Figure 2.2
This sketch has the caption "San Léandro Courthouse, Alameda Co. as Left by the Earthquake of Oct. 21, 1868," and is included in Joseph Armstrong Baird Jr., *California's Pictorial Letter Sheets: 1849–1869* (San Francisco: David Magee, 1967), item 253. Image: Photo 48052:152, Huntington Library, San Marino, California.

And Hector W. Vaughan, a photographer, advertised "thousands of views of the effects of the earthquake … come at once."[99]

The enterprising *San Francisco Daily Morning Chronicle* began advertising a "special edition" that would be illustrated with sketches and published a week after the earthquake. The special edition would purportedly contain "a thorough, reliable and complete history of the great disaster" meant to "make it more intelligible to Eastern and other readers."[100] In the published version, the first page was almost covered entirely with illustrations of the most visually damaged buildings in San Francisco.

Photographers seized the opportunity to profit from the earthquake. Immediately after the earthquake, newspapers anticipated "numerous photographs of the ruins caused by the earthquake."[101] The *Daily Alta California* teased, "The enterprising photographers must hurry up" because "in two weeks more nearly all the damages will have been repaired, and in two

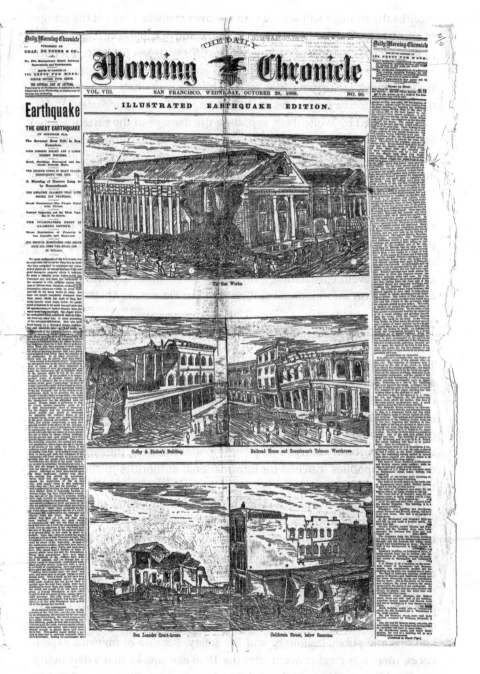

Figure 2.3

Illustrated Earthquake Edition, *Daily Morning Chronicle*, October 28, 1868. Image: Alice Phelan Library, Society of California Pioneers, San Francisco, California.

months the stranger will seek in vain for any extensive traces of the ravages of the greatest earthquake that ever shook and startled San Francisco."[102] Much like the chamber of commerce concerns with San Francisco's reputation, Bay Area newspapers were anxious about how artists would portray the earthquake, worrying that "the publication of their engravings will damage California in the eyes of those living in the Atlantic States."[103]

As explored by Wollenberg, reporting the damage in the eastern United States exposed concerns about how California, as "earthquake country," would be perceived. The San Francisco Chamber of Commerce claimed the authority to tell the story of the earthquake via telegraph to the eastern United States; nonetheless, the narrative did not widely resonate or define the narrative. This downplaying of damage was contested by other telegrams, letters from local observers, imagery on letterheads and in newspapers, and editors of newspapers in nearby cities. With the backdrop of the struggles around the reach of public information infrastructure and how this shaped event epistemology, the rest of this chapter explores the issues associated with informational authority in San Francisco.

The Official Response

The second half of this chapter focuses on the production of information about the earthquake from the local government and California Academy of Sciences. Newspapers encouraged institutions to produce authoritative narratives about both the earthquake itself and how to respond to earthquakes generally. In this sense, the local government and scientific experts had the opportunity to try to shape how to make sense of the earthquake as earthquake publics searched for informational authorities in the information order.

People living in damaged areas agitated for and circulated "official" or authoritative information about the earthquake damage from government officials as well as seismic and building experts. These authoritative documents were publicly circulated in newspapers and printed circulars, and offer insights into the government's limited work in documenting the earthquake. Government officials, such as the mayor of Oakland or members of the San Francisco Board of Supervisors, used the local newspapers to communicate with the public about relocated government buildings, the fate of some public buildings, and fire safety. Because of previous experiences, there was great concern after the 1868 earthquake that a destructive fire would start; thus, much of the early communication was related to fire safety.[104] The *Oakland Daily News* printed an *Extra* on the day of the earthquake with the headline, "Orders of the Mayor!!" that explained fire safety precautions.[105] The next day, the *Oakland Daily Transcript* reported that the mayor of Oakland, Samuel Merritt, had circulars about fire safety distributed throughout the city.[106] The government in San Francisco used newspapers to communicate with the public about fire safety as well, but unfortunately this communication still did not entirely prevent fires.[107]

Newspaper companies also looked to the government for an assessment of buildings. Firewalls were a particularly contentious topic. After San Francisco burned several times in its first decade of existence, firewalls had been erected as a way of containing conflagrations. In the face of earthquakes, however, firewalls were unreinforced piles of bricks waiting to be knocked over. The journals and newspapers complained about these firewalls, and wondered who was responsible. The *Overland Monthly* wrote the following:

> A fire-wall fell ... burying two innocent victims beneath its fragments. Who is responsible? Are republican cities without government, or is that government only for commercial purposes? It were better for the credit of our city that half the brick structures in town should be pulled down than one should fall in another convulsion, burying one invaluable life in its ruins.[108]

Newspapers argued that the government should regulate firewalls and extended this assertion to all buildings in San Francisco. The *Daily Alta* characterized the activities of those who built the buildings that fell: "*Every fatal casualty was the result of criminal carelessness*" [109] The *Chronicle* also used the descriptor "criminal carelessness" to capture the reconstruction process, complaining that "the authorities" ought to take action.[110] The *Daily Morning Chronicle* outlined how it thought the San Francisco Board of Supervisors should become involved through the appointment of a committee of expert architects and builders; it advocated that this committee was not just a way to ensure the safety of residents but also the business interests of the city.[111] While the *Daily Morning Chronicle* advocated for a commission to inspect the buildings in San Francisco following the earthquake, the *Daily Alta California* and *Overland Monthly* wanted a longer-term solution: a permanent position to oversee the proper building practices, superintendent of buildings, or permanent commission.[112] Even though antiseismic building codes would not be addressed in California at the state government level for another seventy-five years, these calls helped set the tone for the expectations of the government's role in earthquake safety.[113] The call for building oversight seems to be supported not only by the earthquake public but also by some newspapers editors who maintained that regulation could make San Francisco actually more appealing to investors—a contrast to the idea that damage should be downplayed for the sake of attracting investors.

To some extent, the discussion about the safety of San Francisco's schools provides an example of how the creation and public circulation of a formal report could satisfy the earthquake public. Many students were at school the morning of the earthquake and were sent home by the superintendent, leaving San Franciscans speculating about when students should return to schools.[114] An architect was appointed by the superintendent and made an official report to the Board of Education on October 24, 1868, saying the school buildings were safe for students.[115] The report concluded with a statement that articulated what the public must have wanted to hear in terms of a guarantee of safety: "I am able thus to place upon record the proof of the undoubted security of all the said school buildings," and the architect's report was printed in several local newspapers.[116] As with the communication about fire codes, the newspapers printed the message from the superintendent to tell San Franciscans about when schools would be open as well as the text of the report by the architect.[117] Regardless of the soundness of these expert building assessments, the public information infrastructure facilitated the circulation of these reports on buildings. In this case, putting the school report "on record" gave the proceedings a level of accountability, asserted informational authority, and settled the discussion over the safety of schools for the earthquake public.

Yet the public circulation of documents did not put an end to the debates over the fates of other buildings. For instance, reports about what was to be done with the customhouse were conflicting, as reports with different results circulated. The first telegrams to the eastern United States said that the customhouse was unsafe.[118] At one point, two members of the US Engineer Corps supposedly "made a survey" of the building, recommended it should be demolished, and telegraphed the report to the Department of the Treasury in Washington, DC.[119] Later, a newspaper article said that the engineers' report was less decisive and that the building was unfit for occupation.[120] Eventually, one of the engineers present at the investigation came forward and stated that he had not even inspected the building.[121] The *Bulletin* invited the public to decide for itself if the structure was affected and attempted to use the confusion about the customhouse to appeal to the populace to dismiss earthquake damage: "It is now believed by hundreds of person who have examined the Custom House building, that it has been but slightly injured, if at all, by the late shock. Thus, another heavy estimate charged to the recent earthquake disappears."[122] This type of appeal to the public's eye in the midst of a confusing debate among experts served the probusiness *Bulletin*'s information agenda of dismissing the earthquake damage. Nevertheless, as the calls for building inspectors indicate, earthquake publics were not always satisfied with the outcomes of these struggles by the newspapers and government to establish informational authority.

Producing Authoritative Information

The earthquake public also waited for an official report on the scientific aspects of the earthquake. These results did not appear in a public forum like the newspapers where the reports of buildings appeared. In order to understand the earthquake, California residents had to confront an uncomfortable tension: clearly people knew that earthquakes were phenomena they had to live with, but they were not well-understood phenomena. There were a number of ways in which authors and editors of newspapers and other periodicals sought to reassure readers—and investors—and make sense of the earthquake: equating earthquakes to other natural disasters experienced in eastern states, arguing for a nihilistic attitude (i.e., the planet earth is one unavoidable disaster), insisting the 1868 earthquake would be the worst earthquake anyone would experience in California, and connecting California earthquakes to worldwide phenomena.[123] Some reports said people ran into the streets screaming because they were reminded of the news of destructive earthquakes in South America that had claimed thousands of lives, but comparisons to horrific earthquakes elsewhere also served to minimize the damage in San Francisco.[124] The *Daily Alta California* reassured readers by describing how similar devastation in other cities after earthquakes could not happen in San Francisco because, for example, houses in Quito, Ecuador, used mud rather than mortar.[125] Discussions of the earthquake acknowledged that *"California is an earthquake country,"* but only so far as to say that it was not like other earthquake countries.[126]

In light of the uncertainty surrounding earthquakes in San Francisco, newspapers as far away as Chicago anticipated the delivery of a report about the earthquake from a group of "learned men" at the California Academy of Sciences.[127] This was not unusual. The *Proceedings* of the California Academy of Science indicate that Dr. John Boardman Trask frequently wrote scientific papers about earthquakes; between 1856 and 1865, he regularly published catalogs of California earthquakes.[128] The *Meeting Minutes* of the California Academy of Sciences reflect that at the first meeting after the earthquake on November 2, 1868, the members thought the public expected that a report would be released. The group resolved to produce a report on the earthquake but "laid over" the resolution because there was not a "quorum present."[129] At the next biweekly meeting of the academy on November 16,

1868, it appears that some progress was made, as one member produced "a specimen ... from the fissure at Hayward, caused by the earthquake."[130] According to the *Meeting Minutes*, "A discussion of the recent earthquake followed and a partial report of investigations was made." At a follow-up meeting on December 21, 1868, Dr. James Blake, vice president of the California Academy of Sciences, presented a map showing "the directions in which the earthquake of October 21 struck."[131] Blake gave his "partial verbal report of his observations on the subject." The *Meeting Minutes* conclude that "investigations on the subject are still in progress."[132] It appears that at the time, the academy members were still gathering evidence to be used to make a report, although a report that fulfilled the expectations of the newspapers and scientists was never published.

Reporting out to the public from the scientific community about the earthquake was also the province of the Earthquake Committee, convened in late November by the San Francisco Chamber of Commerce. This is, of course, the same chamber of commerce that sent the telegraph downplaying the earthquake. For the chamber, the production of a report about the earthquake was not in contradiction with its project of alleviating investor concern but rather was part of it: "This [telegram] fortunately, had the effect of restoring confidence as the present and future solvency and welfare of this community, an opinion that was further heightened, when it became known that its merchants had set on foot an investigation having for its object the matters previously described."[133] Despite the formation of the Earthquake Committee, however, the group failed to produce a significant report. Many scholars who have covered this failure surmise that the report was repressed, there was not enough scientific expertise in California to create such a report, it was never written because the organizer died, or it was underfunded.[134]

I find it valuable to revisit the fated nonreport as another example of how the information order produced a particular epistemological understanding about the earthquake. The Earthquake Committee aimed to do some of what the newspapers agitated for—namely, convene a group of people who would come up with practical advice about building in the future.[135] The subcommittee charged with scientific inquiry included Blake as well as Trask, who had reported on the 1857 earthquake to the California Academy of Sciences and had completed numerous other earthquake investigations. At the annual chamber of commerce meeting on May 12, 1869, the Earthquake Committee reported that while the official report would not be available for two more months, Thomas Rowlandson, the former secretary of the Earthquake Committee, had independently released a pamphlet.[136] The Rowlandson pamphlet did not provide analysis that the California press found relevant. As the *Daily Alta* wrote, "We looked with interest through the new essay, hoping to find something applicable to the wants of Californian builders, but we were disappointed, for nearly all the space is devoted to points that are of very little practical value."[137] Unfortunately, Rowlandson had actually been dismissed from the Earthquake Committee, which he mocked for its lack of seismic knowledge, and was disallowed from using the earthquake reports that the committee had collected, noting in his pamphlet, "At the time of dismissal, it was intimated to me that the communications and other documents accumulated during the proceedings of the Committee were its [the committee's] sole property, and ought not to be used, excepting for its purposes."[138]

These records were unlikely to have been taken by people with a lot of experience making amateur earthquake records, or what Deborah Coen calls "earthquake observers."[139] Blake, in his 1870 report to the chamber of commerce, was dismissive of the work of these observers:

> The plan adopted for carrying out our object was to address circulars, containing certain questions to persons residing in different part of the State, requesting answers to be returned to the Committee. The information obtained by this means was of so vague and often of so very contradictory a nature, that it was worth but little, and it was then decided to employ competent paid agents to travel over different parts of the country where the shock had been felt, to collect what reliable facts could be obtained.[140]

Though the earthquake public seemed interested in supplying scientists with observations about the earthquake, it seems that the scientists could not make use of them or were forbidden from doing so.

In spite of the death of George Gordon, head of the Earthquake Committee, in May 1869, and the pamphlet produced by Rowlandson, the California press was still interested in learning of the projects outcome.[141] But when the report to the chamber of commerce was eventually made in 1870 by Blake and Washington Bartlett (secretary of the chamber of commerce), its contents were not printed in local newspapers alongside notices of the chamber's meeting minutes, as previous updates had been, though the report was published with the *Annual Reports of the Chamber of Commerce*.[142] The *Annual Reports* makes it clear that the chamber had expected to be able to raise money for the Earthquake Committee, and the chamber itself had actually paid for various reports to be written by committee members or other experts, but had not physically acquired them: "As these [reports] were obtained at a heavy expense, it is to be hoped that any one who may be unwittingly withholding them, will return them to the Secretary of Commerce."[143] The underfunding of the project and Gordon's death resulted in loss of existing documentation and an inability to fund further work. There was not the political will to raise the funds to complete the report. The production of this document was a promise made to assuage nervous investors, not something that the chamber of commerce actually wanted to create.

Though the chamber lacked the political will to complete the report, the earthquake public did not seem to lose interest in the earthquake; two months after the earthquake, the *Daily Alta California* published a letter saying, "It is natural to suppose that more reliable information can be brought together in this State, both of practical as well as theoretical nature."[144] Despite the fact that the earthquake public did not get the official report it sought (from either the Academy of Sciences or Chamber of Commerce Earthquake Committee), it still made sense of the earthquake and learned from it. The San Francisco newspapers printed "lessons" from the earthquake, many of them authored by people working on the Earthquake Committee, or with the local government on inspections of public buildings.[145] The evening of the earthquake, the *Bulletin* wrote about "the Earthquake and its Lesson."[146] The lessons of the earthquake were often tempered by an attitude that implied that sturdy building practices were obvious and the solution to avoiding future damage was simply a matter of people engaging in the best building practices.[147] In fact in a few cases, the press adopted a nearly Darwinian view of buildings after earthquakes—namely, the weak buildings were destroyed in the earthquake, leaving the city filled with strong buildings—but this was not the majority of stories.[148] The published lessons were apparently so pervasive that the *San Francisco Daily Morning Chronicle* said, "Since Wednesday last we have been treated by the city press to no less than seven long-winded, tedious, prosy disquisitions on 'The Lessons of the Earthquake.'"[149] The challenge for many San Francisco residents, who had moved from all over the world, was to understand how to adapt to California earthquakes—not something familiar to new immigrants. *Scientific American*, printed in New York, summarized this attitude in an article titled "The California Earthquakes—A Different System of Building Necessary."[150]

Without scientific explanations of the earthquake or the government acting as an informational authority, people were faced with a conflicting cacophony of stories undoubtedly shaped by business interests in San Francisco that were promoting California, and by people interested in selling newspapers and other periodicals. In the midst of these conflicting narratives, earthquake publics also shared construction and public safety information in print through lessons learned, articulating informal knowledge.

Conclusion

All Californians were involved in the tasks set forth by Knight in his letter to his mother: evaluating "exaggerated" accounts versus reports that "make light of the whole affair." Earthquake publics in damaged areas could see what happened to their communities with their own eyes, but they were still trying to sort out the fate of other locales. Without trusted institutions to tell a story of the earthquake, most people were left, like Knight's family, trying to make sense of a multitude of accounts of a frightening event.

This chapter examined how the reach of public information infrastructure and "instant" connection of the telegraph shaped reports about the disaster. Most personal letters from residents to people located outside California presumed, as Knight's did, that they were writing in the wake of a telegraphed version of the story, and indicated some anxiety about the time between when the first news of the earthquake arrived via telegram in the newspapers and when their own letters arrived in the mail.[151] These tensions around the delivery of different versions of the story—ones that exaggerated and made light of the earthquake—built up not only in personal relationships but also on a political level. The telegraph infrastructure made it possible for some stories to get through to the eastern United States immediately, while stories from others such as Knight, who possibly could not afford to send a telegram, were sent by mail.[152] Reports about the earthquake and images of disaster were created with this vast infrastructure in mind. Sometimes this meant that organizations such as the chamber of commerce wanted to downplay the earthquake for investors. On the other hand, printers or newspapers such as the *San Francisco Daily Morning Chronicle* wanted to sell lots of papers and printed the most sensational images of damage. Many of the reports produced in the days after the earthquake were in the heat of emotional moments of concern.

Through the production of the public information infrastructure, the people of California wanted to understand what had happened, how broad the reach of the earthquake was, and what could be done. The local government marshaled some expertise and reporting practices to document the current state and future of government-owned buildings. But it appears that the earthquake public wanted more guidance on privately owned property as well. It also points to issues in the practice of producing the kind of reports that were in demand. The Earthquake Committee, a group of scientists, architects, and engineers, had the opportunity to establish some kind of informational authority, and thus explain to people what had happened and how to establish safe building practices in the future, but it did not produce the definitive document that the earthquake public sought.

Public information infrastructure was both an instrument or source for interpreting what happened and a means for producing and circulating reports. Working and nonworking physical infrastructure served as a way for people to assess what happened, though not necessarily with any accuracy. The information order enabled the production and circulation of particular documents to explain what happened, including reports about damage in local areas, inaccurate aggregated estimates of damage, rumors of nonworking infrastructure, imagery of damaged property, assessments of government buildings, and lessons for rebuilding. Authoritative independent expert analyses, however, are notably missing from this information order.

In these moments there appears to be a kind of imbalance between what the public hoped to learn and what documents were actually being circulated. How do we account for the constitutive qualities of public information infrastructures when something that is in demand is not produced, like the authoritative report? This case can be indicative of powerful capitalist forces that wanted to suppress extended discussions about the costs or dangers of earthquakes. But it also points us to the fact that the production of formal reporting is an expensive process. Those charged with producing the Earthquake Committee report raised repeated calls for funding and underscored the lack of its availability. This kind of formal reporting requires a combination of expertise, bureaucratic experience, political will, and established funding channels not present in the information order of California in 1868. Even without a formal report, earthquake publics, both those that experienced the earthquake and those far away, still made the event epistemology through a variety of public information infrastructures.

3   Accounting for People after the 1906 Earthquake

I went directly to the W. U. [Western Union] Office which was a wreck. However, there were hundreds ahead of us and we worked our way through the debris to the desk. When I saw the pile of telegrams waiting to be sent and was told that the wires were all down I left the office at Pine & Montgomery & went to the Postal [Postal Telegraph-Cable Co.] at Montgomery & Market. The office was dreadfully wrecked but one machine was ticking away so I left my message. When the fire swept all away I thought that possibly all messages were destroyed. The next day I sent a message by W. U., a young man who was going to Haywards ... took them ... to send them.[1]

During and after the 1906 San Francisco earthquake and fire, correspondence between Sarah Phillips of San Francisco and her fiancé, George W. Jones of Schenectady, New York, was dominated by discussions of how to best get in touch, how the telegraph system was working, and when and how George found out about the earthquake as well as Sarah's well-being.[2] Many San Francisco earthquake survivors had similar interactions with friends and relatives near and far in the aftermath of the April 18, 1906, magnitude 8.0 earthquake that shook buildings in San Francisco to the ground, breaking many of the pipes that carried water and gas in the city. Fires that started as a result of the earthquake raged for four days, leaving approximately half the city's population homeless and destroying at least two-thirds of the built-up area, including the business district. The disaster was immediately documented and retold in a number of books, magazine articles, photographs, and even theater productions. Personal correspondence also provided critical insights into the event. But the information order in 1906 was different than that of 1868; the population of San Francisco had more than doubled, major newspapers had huge circulations, telegraphy had grown, and the telephone had been introduced.[3] The wealthiest San Franciscans owned the city's newspaper companies, which occupied the most iconic downtown San Francisco buildings and were central in the San Franciscan public sphere. Much like the chamber of commerce in the 1868 earthquake, however, many of the 1906 newspapers participated in a campaign to downplay the earthquake's severity.

The day of the earthquake, Sarah wrote to George, "I have written and wired you but I do not believe you will receive either message."[4] That same day, George wrote to Sarah that he "read every bit of news that has come in the papers," telling her, "I am so anxious to hear."[5] Sarah understood the magnitude of George's emotional burden, and in a later letter sought to comfort him: "I know how you felt. ... And that is the reason I hurried to assure you of our safety."[6] The means for instant news, the telegraph, which Sarah and George relied on, didn't work after the earthquake and fire.

Whereas most of Sarah's telegrams to George were lost, most likely in the fire, on April 28, almost a week after the earthquake, Sarah received two telegrams from George by mail.[7] Sarah sent regular letters with updates about her ordeal and augmented her letters to George with clippings from San Francisco newspapers.[8] The mail system, Sarah reported, "allows all mail to go through without stamps or envelopes and such funny things go into the box."[9] But the task of connecting mail with the correct recipients was not always easy. Sarah lost her apartment in the fire, and once the fire finally stopped, she and her companion walked over a hundred blocks around hilly San Francisco to find their mail—which had been sorted and was waiting for them at Sacramento and Fillmore.[10] People across the city were trying to locate their friends and loved ones. Sarah reported, "We are trying to locate our friends. ... [I]t is hard to see mothers looking for children and husbands for wives." She described the ways in which displaced San Franciscans attempted to find each other: "Cards are tacked all over fences, poles, etc. Asking different ones to report at certain places."[11] Sarah worried about a friend she had not heard from, Lizzie Gleason, and advertised for her, wrote letters, and was planning on going to all the registries to look for her.[12] Registries generally attempted to collect and organize records describing the whereabouts of displaced earthquake survivors.

After news of the earthquake, people were desperate to hear from friends and loved ones, but also employees, employers, and those who owed them money. Because the earthquake and fire destroyed so much of San Francisco, thousands of people scattered all over the Bay Area, and each person's social geography shifted. Friends, family, and places of work were suddenly in different places. For people who did not own land, this relocation was possibly permanent. People updated each other with their new locations by telegram, if the telegraph was working, or mail, if they knew where to send the letters.

The most extreme cases of people attempting to locate loved ones involved physically going into San Francisco to find them. Stories circulated of people searching for loved ones by stuffing notes inside loaves of bread destined for refugees.[13] As Sarah described it, the city of San Francisco was littered with notes of people attempting to find each other. Residents of San Francisco designated a fence on which loved ones might leave notes as they attempted to locate each other.[14] Bulletin boards near where refugees fled also were places people could physically go.[15] These symbolic traces of people's whereabouts dotted San Francisco, as signals that life would come back. Photographs of the burned areas of San Francisco show painted bedsheets and other handmade signs hanging from the wreckage, proclaiming that a business would return.[16] The system of posting notes and signs had limitations, though; people were so spread out that there was no guarantee anyone would happen on a sign. Registration bureaus cropped up as the form taken by many efforts at putting all these signs and notes into more recognizable repositories. Newspapers, fraternal organizations, relief committees, and the police all set up registry bureaus around San Francisco and the surrounding areas to which people had fled. Still, these different bureaus weren't necessarily coordinated; Sarah explained that she might go to *all* the registration bureaus to look for a friend of hers. Furthermore, many people put out notices in newspapers and other publications inquiring about the status of others or giving their own whereabouts. These notices and methods of accounting for others give researchers like me a view into how the earthquake publics were constituted through the public information infrastructures.

The earthquake and resulting fire destroyed significant parts of the public information infrastructure in San Francisco, making telegraph lines inoperable, ruining telephone exchanges, and burning printing presses and paper. It also scattered San Franciscans throughout the Bay Area. If we understand that public information infrastructure is an ongoing process, it needs to be asked, What happens when there is major destruction such as the April 18, 1906, earthquake and subsequent four-day fire? What is so amazing, in some sense, about the 1906 earthquake and fire is that the process of infrastructure, despite all the horrific damage, continued in some form. And it is the stability of social, political, and economic forces that enabled the continuity of public information infrastructure along with, in some sense, San Francisco. In reconstituting the public information infrastructures, old institutions adopted new techniques and new institutions adopted old techniques. This points toward a story of the continuity of the public information infrastructure after a disaster. The story of remaking the

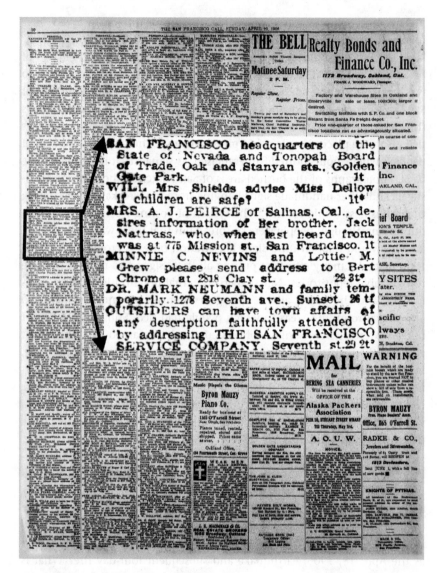

Figure 3.1
Personal advertisements in San Francisco newspapers showed people trying to get in touch with each other. *San Francisco Call*, April 29, 1906. Image: California Digital Newspaper Collection, Center for Bibliographic Studies and Research, University of California at Riverside.

public information infrastructure is one in which the way that people organized and worked proved powerful. Where physical telegraph infrastructure faltered, the bureaucratic work practices of the post office remained.
Despite the destruction of the newspaper presses, newspapers remained
the best way to quickly broadcast personal news—a powerful position. The
relief committee registration process represented new institutions for citizens to deal with, but these processes were built on the familiar practices of
the census and Progressive era charity rehabilitation ideologies.

This chapter is about how earthquake publics were accounted for after
the 1906 earthquake—essentially, how the public information infrastructures of 1906 helped and didn't help Sarah as she attempted to notify
George and locate Lizzie. People affected by the earthquake were accounted
for in their personal relationships, by the post office trying to deliver them
mail, registration systems attempting to account for their whereabouts,
charitable organizations attempting to serve them, and people—in 1906
and today—wanting to know who died in the earthquake and fire. This
event epistemology reveals the ways that some, but not all, earthquake
publics might have been visible. While some people found their loved ones
by physically going and looking for them, for many others, particularly
those who were far away from the earthquake and wanting to know what
had happened to friends and family, this search was an intensely informational activity. The accounting was actuated in and through the public
information infrastructure. In a literal sense, the denizens affected by the
earthquake, the earthquake public that needed to be accounted for, were
constituted in and through the public information infrastructure. But as
the last chapter argued, this public information infrastructure was social,
political, and economic, and these forces were brought to bear by the people, technologies, and institutions that made the public information infrastructure in their everyday practice.

Prior scholarly writings about the 1906 earthquake have not put information orders at the center of study, yet as Sarah and George's story makes
clear, the public information infrastructures used to account for one
another were central to many people's early experiences of the disaster. I
do not wish to imply here that public information practices are the most
important analytic one can apply to studying disaster, but they provide
one more window into the experience of the earthquake for San Franciscans. I build on scholarship focused on how the earthquake generally was
portrayed to serve particular economic and political agendas. Researchers
have described how newspapers colluded with the railroad giant Southern Pacific and others to ensure a "seismic denial" narrative of the disaster

that revolved around the fire while ignoring the still politically inconvenient earthquake.[17] Yet many books, plays, photographs, and other media published in the aftermath of the earthquake did not participate in downplaying the earthquake, but did exactly the opposite, sensationalizing the disaster.[18] In addition to exploring the information infrastructure practices, I build on the more recent scholarship of historians Andrea Rees Davies and Marion Moser Jones, who describe the practices of the official disaster relief organizations, all steeped in the Progressive era "scientific" charity ideals of the time.[19]

Notification with Telegrams and Letters

Personal Use of the Telegraph

The telegraph was in demand after the earthquake and fire, particularly outside San Francisco. People all over California and the United Stated mobbed telegraph offices attempting to send telegrams to San Franciscans.[20] In a 1919 set of recollections of the earthquake by University of California at Berkeley students, who were young children living outside the affected area in 1906, many students remember their families finding out about the earthquake in a newspaper and then anxiously sending telegrams to get in touch with family members inside San Francisco.[21] "There was considerable excitement in the store as everyone gathered to read the paper. ... There was also considerable anxiety in our home during the ensuing days because of the difficulty in getting a reply to a telegram to some of our relatives who were visiting in San Francisco at the time of the fatal calamity."[22] That replies were not forthcoming was a source of anxiety for many families: "I found my folks already preparing to send a telegram in order to inquire about the fate of relatives living in San Francisco. We waited impatiently for three days before definite word of their safety reached us."[23] In one case, it was not the telegram that relieved the family's anxiety but instead a letter: "I remember my mother's sending numerous telegrams to relatives. ... After a few days we received some mail and were greatly relieved. ... [P]ieces of cardboard without stamps and the postage was paid when they reached us."[24] Messages were often simply, "Alive and well; lost everything."[25]

Even the world's most powerful were at the mercy of the broken and backlogged telegraph network. Communications from international diplomats to the secretary of state sought personal information about loved ones in the affected areas. Heads of state with connections to people in the Bay Area included inquiries into the well-being of specific individuals in their official correspondence with the US Department of Defense. A telegram addressed to "The Governor of California, Sacramento," stated, "At the insistence of Austro-Hungarian Ambassador, I have the honor to request information regarding fate or whereabouts of Charles Siegler, whose address is given as 'General Delivery, San Jose' and of Hugo Bettelheim, who is described as 'Merchant at Burlingame near San Mateo.'" A handwritten reply reads, "Referring to inquiry Charles Siegler native Austria Province Bohemia Alive and well," and is signed, "Geo. C. Pardee, Governor."[26] But most people did not get responses from the California governor as to the whereabouts of the people they sought; the majority was stuck with the overwhelmed telegraph wires.

Sarah and George were dependent on the telegraph companies—Western Union, Pacific Telegraph and Telephone, and the Postal Telegraph Cable Company—all of which Sarah tried to use the day of the earthquake. In San Francisco, for several hours after the earthquake, the Postal Telegraph Cable Company was operating and hence able to give news about the fire's progression. The fire eventually shut this office down, making it impossible for people in San Francisco to telegraph people outside the city.[27] Afterward, the manager of the Postal Telegraph Cable Company was quoted as saying that the staff bravely "worked in San Francisco during the fire until we were put out of the building."[28]

Accounts of telegraph operators and telegraph office managers describe the incredible effort of these workers to find telegraph lines in San Francisco and Oakland that had connections to the outside world. After the earthquake occurred at 5:12 a.m., the main San Francisco office downtown had four working wires to Chicago and one to Los Angeles. Postal Telegraph operator Mr. A. J. Esken made it to the main San Francisco office about an hour after the earthquake. Esken used his wire primarily for transmitting dictation from a newspaper reporter. Many of the other Postal Telegraph office employees, including the electrician, rushed to work to make repairs and assist with the intuited onslaught of telegrams. As the fire spread, the operators were required to evacuate their building at 11:15 a.m., at which point Esken grabbed his typewriter and fled. The office's generator, a Dynamo, was also saved. Over the next few days, Esken, like other men in San Francisco, was conscripted to carry a fire hose and then added to a citizens' patrol. Three days later he made it to the Oakland office of the Postal Telegraph, where he continued to work.[29]

The difficult work of ensuring the continuation of public information infrastructure fell to the telegraph company employees who worked in difficult conditions to come up with creative work-arounds to the broken physical infrastructure. The superintendent of the western division of the

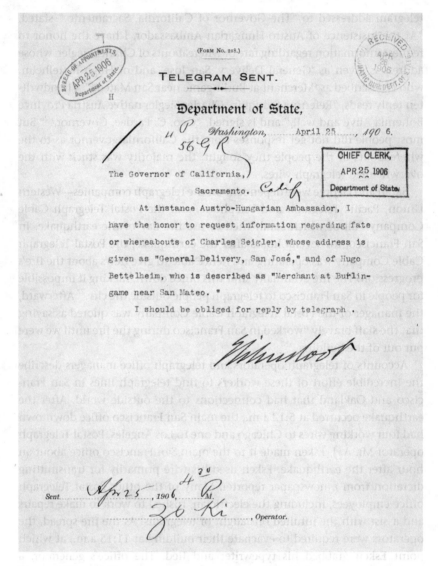

(FORM No. 218.)

TELEGRAM SENT.

II

Department of State,

Washington, ...April 25...... , *190* 6.

The Governor of California,
               Sacramento.

CHIEF CLERK,
APR 25 1906
Department of State.

At instance Austro-Hungarian Ambassador, I
have the honor to request information regarding fate
or whereabouts of Charles Seigler, whose address is
given as "General Delivery, San José," and of Hugo
Bettelheim, who is described as "Merchant at Burlingame near San Mateo. "
          I should be obliged for reply by telegraph .

Sent......*Apr 25* ....., *1906*,.........*M.*

.........................., *Operator.*

Figures 3.2a and 3.2b
Diplomats inquired about the status of potential earthquake victims and received
personal responses from the governor of California, George Pardee. Image: Depart-
ment of State to Governor of California, April 25, 1906, and George C. Pardee, Gov-
ernor of California to Secretary of State, April 29, 1906, box 1, NARS A-1, entry 182;
Official Messages on the San Francisco Earthquake, April 19–25, 1906; Messages of
Condolence; Special Series of Domestic and Miscellaneous Messages of Condolence;
Miscellaneous Correspondence, 1784–1906; General Records of the Department of
State, Record Group 59; National Archives, College Park, Maryland.

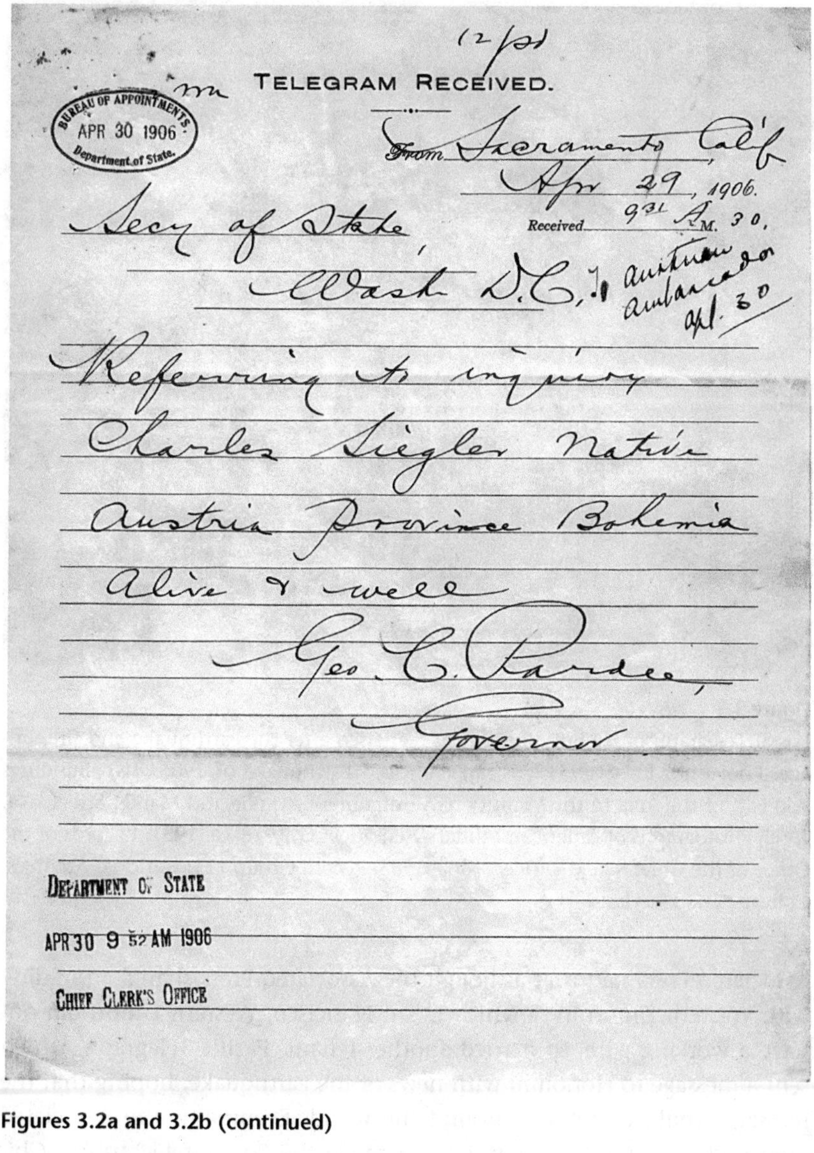

TELEGRAM RECEIVED.

(2 p)

BUREAU OF APPOINTMENTS
APR 30 1906
Department of State.

From Sacramento Calif

Apr 29, 1906.

Secy of State,

Received 9³¹ A.M. 30,

Wash D.C., Austrian ambassador apl. 30

Referring to inquiry Charles Siegler Native Austria Province Bohemia Alive & well

Geo. C. Pardee,

Governor

DEPARTMENT OF STATE

APR 30 9 52 AM 1906

CHIEF CLERK'S OFFICE

Figures 3.2a and 3.2b (continued)

Figure 3.3
Postal Telegraph office building in San Francisco after the earthquake and fire. Image: Photo no. 111-SC-95177; "San Francisco Earthquake of 1906: Two Buildings and Part of the Area in the Vicinity of Montgomery Avenue and Market Street," ca. 1906; Photographs of American Military Activities, ca. 1918–ca. 1981; Records of the Office of the Chief Signal Officer, 1860–1985, Record Group 111; National Archives, College Park, Maryland.

Associated Press said that although the Associated Press shared a building with Western Union in downtown San Francisco, Western Union did not have a working wire, so it tried another tenant, Pacific Telegraph, which sent a message to Honolulu with news of the earthquake, hoping that the message would be relayed around the world. Eventually, the Associated Press ended up sending news on a wire via the Postal Telegraph to Chicago.[30] Western Union officials claimed that the first story of the earthquake was sent "by means of relays, Los Angeles, Salt Lake, Denver and other places" by Western Union chief operator H. J. Jeffs, perched on a telegraph pole in Oakland.[31]

As Jeffs depicted the state of the telegraphic infrastructure in Oakland, "The wires were crossed in a hundred places, broken and grounded."[32] The morning of the earthquake, Jeffs was able to find a connection between West Oakland and Sacramento. He began setting up a makeshift office in West Oakland where the cables emerged from the water.[33] Three days after the earthquake, a Western Union electrician arrived from Chicago via train with new telegraphic equipment to set up the new Western Union office in West Oakland, which was quickly connected to offices beyond Sacramento.[34]

Overall, telegraph companies were wealthy and deployed their vast resources to quickly repair parts of public information infrastructures—parts crucial for people to alert loved ones about their well-being. But the telegraph wires were far from operating as they did prior to the earthquake, and companies such as Western Union often couldn't handle the volume of telegraphy traffic. A Western Union official commented on the "vast amount of business," noting "the magnitude of which never before was experienced on the Pacific Coast."[35] Even though its telegraph cables were damaged and mostly nonfunctional, Western Union continued to collect telegrams throughout the afternoon of the day of the earthquake.[36]

The telegraph was critical to the process of news collection for the newspaper companies, and they described their work-arounds extensively.[37] One newspaper dramatically claimed that San Francisco was without service "for the first time in history."[38] That the newspapers and telegraph were completely intertwined in their operation became clear as the nonworking cables were the subject of several stories. Despite the fact that all the daily San Francisco newspapers saw their offices burn to the ground, a *San Francisco Bulletin* editorial declared that "the maintenance of telegraphic communication with the outside world has been one of the most difficult matters with which the newspapers have had to contend."[39] As in 1868, newspaper companies were heavily dependent on the telegraph for stories from far away. The telegraph companies quickly set up offices in Oakland to attempt to temporarily replace the multitude of offices in San Francisco.[40] Interestingly, the only telephone in the area that would have been available for nonmilitary purposes was converted to a telegraph.[41] Makeshift telegraph offices were eventually set up in the burned districts of San Francisco.[42]

Of course, the telegraph companies were not just employed by the newspaper companies; they were also in demand for personal telegraphy. In the Bay Area, many anticipated, as Sarah had, that their loved ones would want to hear from them, given the dramatic news. One San Franciscan reported his experience: "The knowledge of friends and relatives at the mercy of the yellow press sent me quickly to the Telegraph

Figure 3.4

The Western Union office in West Oakland, which was constructed quickly after the earthquake. The original photo caption reads, "New Main Office—completed in less than a week." Image: Western Union Telegraph Company Records, Archives Center, National Museum of American History, Smithsonian Institution, Washington, DC.

office. I stood in line for ten minutes before I wasted my money on messages that were never sent or got lost on the way."[43] As Sarah explained, the telegraph offices in San Francisco on the day of the earthquake had large queues; the same was true in Oakland, and the lines there likely only got longer as hordes of San Franciscans fled there.[44] By all accounts, the lines at the telegraph offices were interminable. Telegraph operators were then left with a lot of discretion about what got sent over the limited connections.[45] Many newspaper stories said that agents dealt with telegrams in the order received, although other articles make it clear that there was much discretion on the part of the operators in deciding whose telegrams were prioritized.[46]

The makeshift telegraph operations being set up in Oakland were crucial sites for people to alert others about their well-being. Yet because telegraphers had a lot of discretion about what got sent, personal messages were not prioritized above newspaper company reports. Outside the Bay Area, people were desperate to get in touch with their contacts in San Francisco and trying to figure out the status of the telegraphic infrastructure. The *Los Angeles Herald* declared in screaming headlines that it was "impossible" to send telegrams.[47] The *Los Angeles Times* claimed it whipped the entire city into a frenzy, releasing six extras throughout the day as every bit of news about San Francisco trickled in via the Postal Telegraph: "The extras were bought with a fever and read with an avidity beyond precedent." At this point, "Thousands of people in Los Angeles had thousands of relatives in San Francisco," and likely wondered, "How would it be with them?" The worried Angelinos "filed" telegrams that were not sent.[48] The *Los Angeles Times* said that newspapers exclusively used the one working Postal Telegraph cable from 8:00 a.m. to when the cable went down. This meant that none of the concerned citizens of Los Angeles could wire friends in San Francisco.[49] Even as newspaper companies printed extras telling of every moment of destruction, people would still hang out around the telegraph offices, waiting for the latest updates about San Francisco.[50] The telegraph services prioritized newspaper companies over individuals on the few available working wires at the same time that the horrible news coming from San Francisco underscored the need to get in touch with contacts there. The telegraph companies claimed that they "served the newspapers of this country for two weeks with all the news we could get free of cost."[51] Telegraph operators took dictations from reporters, keyed ready copy, and wrote about their own experiences and impressions in the reports sent. But readily available news only reinforced the fact that the wires were unavailable for personal use.

Figure 3.5
The caption for the image reads, "Crowds trying to telegraph relatives from Oakland offices." Image: California History Room, California State Library, Sacramento, California.

In the case of Los Angeles, a city with many ties to San Francisco, the disconnection with San Francisco was so problematic, situation so desperate, and backlog of telegrams so great that "the Western Union Telegraph Company sent more than 5,000 private messages via train to San Francisco and Oakland aboard the *Owl* to be delivered by special messengers."[52] Trains from Chicago apparently carried telegrams to the Bay Area as well.[53] Trains also carried telegrams in "great bulk" out of San Francisco.[54] Telegrams were not delivered via the wires but instead were physically transported across the country by boat and train.

Sarah and George's story illustrates how telegrams were sent through the mail.[55] This was not only noteworthy to Sarah and George. The San Francisco Grand Jury issued a report in June 1906, two months after the earthquake, admonishing two telegraph companies in general and Western Union in particular for collecting upward of one million dollars in fees from senders and then sending telegrams through the mail—which as discussed in the next section, was actually a free service provided by the postal service to earthquake survivors. As Colonel Clowry of Western Union explained:

> Of course we mailed messages. We always do in a crisis of this sort. That's nothing new. Whenever there is trouble and we can't get our messages through we telegraph to the nearest point we can reach and then either mail or send special messengers with the messages. There is never a time on this continent that our lines are not down somewhere. Suppose a storm destroys our lines in Iowa. There will be a big accumulation of messages in Chicago, with no possibility of forwarding them under thirty-six hours. Then we would send a special messenger by train to the nearest station beyond the break and telegraph from there.[56]

Yet what seemed like standard procedure for Western Union officials to operate broken infrastructure was largely invisible to others, and when it became known, it was shocking. Newspaper articles describing the Grand Jury investigation noted that San Franciscans and people all over the country were desperate to get in touch with one another.[57] The report conveyed fury that Western Union and the Postal Telegraph Cable Company took people's money and then sent their telegrams through the mail. The telegraph company bosses reacted strongly to these accusations, calling them "unfair" and "absurd," reiterating the sacrifice of the telegraph operators who spent their postearthquake days working rather than attending to their families, and reminding earthquake publics that they sent messages for the newspaper companies and relief organizations for free.[58] In refuting the Grand Jury's charges, the telegraph company managers revealed a great deal about how the telegraph companies actually worked after the earthquake and fire. "When they [telegraph operators] got messages they sent them, frequently by mail, to the place they thought was most likely to reach the receiver."[59] In accepting money for the telegrams and routing them any way they could, the telegraph companies appeared to the San Francisco Grand Jury to be price gouging when people were at their most vulnerable. To the telegraph companies, there was nothing unusual going on; if anything, they believed they were acting benevolently.

Despite the Postal Telegraph and Western Union employees working extra-long hours to repair the telegraphic infrastructure, the number of people trying to send telegrams and, eventually, number of telegrams sent to the Bay Area overwhelmed the companies.[60] Telegrams for people of the Bay Area made their way to telegraph offices to be delivered by messenger, await retrieval, or if the recipient was not locatable, languish.[61] Managing the backlog of telegrams received required that people work day and night to sort through them. It was not just that the telegraph lines were congested and stacks of telegrams needed to be sent; it was that incoming telegrams were practically undeliverable.[62] Prior to the earthquake, telegraph messengers delivered telegrams within cities.[63] But after the earthquake and fire, telegrams took many days to reach their destination because sometimes people could not be found.[64] "Even ... an army of messengers ... would be of little value for the reason that the people are scattered far and wide and a journey from the Ferry Building to Western Addition, or to the refugee camps consumes many hours."[65] Lists of the telegram recipients appeared in local papers, the Oakland telegraph offices, and at the Ferry Building in San Francisco.[66]

The wealthy telegraph companies were happily embedded in the work of the powerful newspaper companies and maintained a powerful position in the information order despite breakdowns. The usually invisible telegraphic infrastructure became visible to ordinary San Franciscans in the course of both the earthquake, as newspapers reported on the various maneuvers of the telegraph companies repairing their lines, and afterward, as the telegraph company bosses explained how they operated during the crisis. Most users of the telegraph were elites, and much of the use of the telegraph was for business purposes.[67] Emergency use of the telegraph after the earthquake likely broadened the potential clientele, yet the telegraph companies favored the newspaper companies and relief organizations run by San Francisco elites over personal messages. A much more familiar technology to working-class San Franciscans, the postal system, was nevertheless attempting to be up and running quickly after the earthquake.

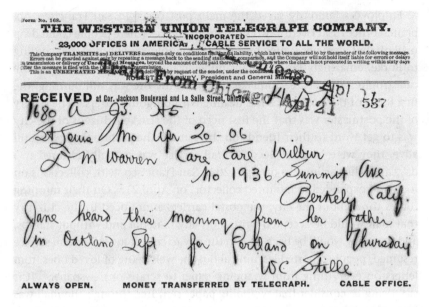

THE WESTERN UNION TELEGRAPH COMPANY.

Figure 3.6
Telegram sent via Western Union showing a stamp stating, "Train, from Chicago, Apl 21." Image: California Historical Society, San Francisco, California.

The Post Office

The fact that telegrams became post office mail was not lost on William Burke, secretary to the San Francisco postmaster at the time of the earthquake. Burke wrote the most celebratory and probably most complete account of the postdisaster post office activities in a twenty-year anniversary issue of the *Argonaut*: "The best proof of the inadequacy of the telegraph service was the fact that thousands of telegrams were being mailed under two-cent stamps in the Post Office, the telegraph companies trusting to the postal service to deliver them."[68] After the earthquake, the post office system was an essential institution of the information order that supported people's efforts to account for each other's well-being. The post office, a comparatively old-fashioned system compared with the telegraph, ended up being the more reliably functioning entity for most people—even the poorest people, who could not afford to send telegrams. It is not to say that the post office was independently functional; it surely relied on telegraphic capabilities to communicate between offices. But how did the post office, whose very existence relies on people being at certain places, regroup so quickly after the earthquake when those places no longer existed?

Employees famously saved the main post office in San Francisco. Six branch offices and twenty-two substations (places where people could drop mail off, but were not staffed by post office employees) burned.[69] According to Burke, amazingly (and perhaps unbelievably), no mail was lost in the catastrophe. After the earthquake and while the fire was still burning, the first focus of the post office was dealing with outgoing letters. "The theory of the postmaster was that the first need of the stricken citizens of the city was to get word to their friends on the outside that, while they were still alive, they were in great need."[70] Mail service was partially restored two days after the earthquake on April 20 in San Francisco, with collections on foot, and expanded to mounted collectors on April 21.[71] On their morning routes throughout the city, the postal carriers announced to the citizenry that mail would be accepted written on any material and without stamps, and that they would be back later in the day to collect it.[72] After mail service resumed, people reported learning about the well-being of loved ones from letters on everything from wrapping paper to scraps of newspaper.[73] It is surprising to consider that wrapping paper sent free through the mail was integral to how some people were represented to their loved ones through public information infrastructures. These choices by the post office as well as other innovations described in more detail later were key to the event epistemologies of many people.

Getting letters out of the city was a priority for the post office, but that seemed almost easy compared with the challenge of delivering mail to residences that no longer existed or people who were scattered. Postmaster Fisk publicly asked that people address letters to San Franciscans to their original addresses, unless they had new, known ones.[74] The delivery effort was further handicapped when 42 clerks and 25 carriers were reassigned from the 350 available in the San Francisco office to Oakland, and 4 clerks to Berkeley, to help with the influx of refugees.[75] It "seriously delayed" the working of the San Francisco post office—"like taking the life preserver from a drowning man and telling him to swim for it."[76] By June 1, 1906, Burke estimated that the post office had almost two hundred thousand "forwarding orders and changes of address." Newspapers and magazines wanted their subscription lists reviewed, and there was duplicative information about businesses. Carriers whose districts were burned worked to keep files on where people had relocated, and eventually district lines were redrawn and new routes established. Overall, the post office reorganized quite quickly to deal with the mail of people who had been "burned out."[77]

The post office's mandate for universal service meant that it included those displaced by the earthquake. Archival photographs indicate that post office stations were set up in refugee camps to reach those displaced.[78] The evidence suggests that either the post office devised methods for reaching into the refugee camps or people took it into their own hands to create post office stations.[79] Either way, having post offices in refugee camps enabled the most disenfranchised people to participate in the United States' oldest information system.

An analysis of postearthquake mail indicates that the post office adopted several mail annotations to deal with the new San Francisco geography. Randy Stehle, postal ephemera collector and historian, describes several markers that helped the post office sort the volumes of mail. First, "burned out" was used as an "auxiliary marking," in which the post office placed a specific rubber stamp on an envelope to indicate the address no longer existed.[80] There was also a "Camp Ingleside" stamp used to forward mail to the refugee camp of that name.[81]

People reportedly wrote to the post office in large numbers inquiring about the whereabouts of their contacts, in hopes that the post office had acquired their forwarding addresses. Yet despite the unprecedented nature of the disaster, the post office adhered to regulations that declared, "No information shall be given by the Post Office regarding addresses." Thus, those asking for the addresses of "obscure parties" likely did not get a prompt reply.[82]

The innovations by the post office also included new methods of keeping track of addresses for forwarding. Initially, the post office used a system of reforwarding, which meant that a letter might be reforwarded several times; eventually a system of shortcuts emerged. Burke, the secretary to Postmaster Fisk, described how the work practices of post office employees evolved, or didn't, after the earthquake. He explained that the system of shortcuts for mail forwarding was not immediately instituted because post office employees continued with their familiar work practices while they were trying to understand the enormity of what had happened and how to approach their new tasks.

It might naturally be asked why [instituting] this [new sorting technique] was not done at once. Because it was impossible, and because it was inadvisable and dangerous in the conditions in which the service was just after the catastrophe. It was impossible, because no one had ever met such a situation before; and it may be said, as soon as we got used to the problem we found a solution for it. It was inadvisable and dangerous because our only hope of maintaining service lay in

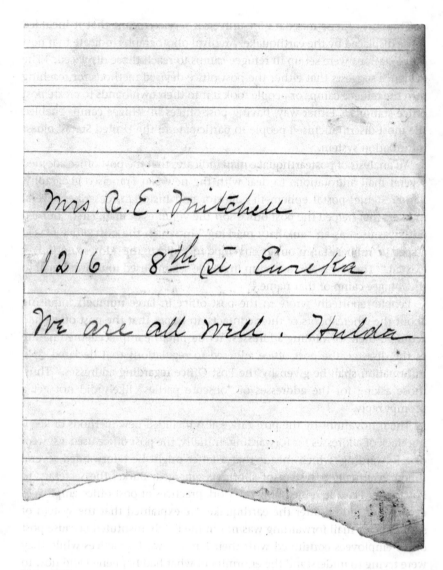

Figure 3.7
A postmarked scrap of paper from April 22, 1906, stating, "We are all well." Image: California Historical Society, San Francisco, California.

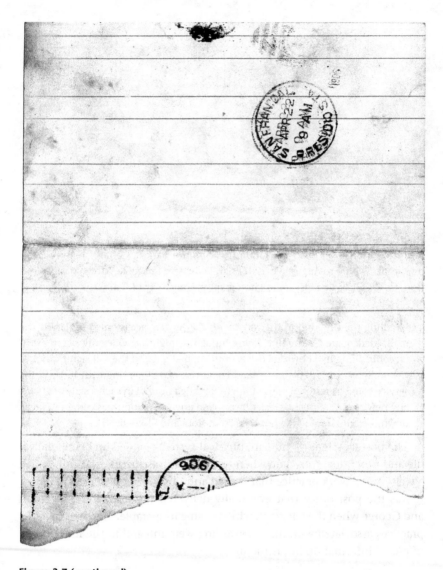

Figure 3.7 (continued)

Figure 3.8
Stereograph of Post Office Substation Q, Hamilton Park (an official refugee camp from June 3, 1906 to August 31, 1907), San Francisco. Image: Keystone-Mast Collection, UCR/California Museum of Photography, University of California at Riverside.

handling the situation in the way in which we were accustomed to handle the mail, inadequate for the time being though it might be. Our only safety lay in creeping along by known methods until we had grasped the full significance of the complicated task before us, and could risk experimenting with it. Until the service could be restored to what might be called normal in such a state of affairs, any new plan that might break down,—and any innovation was certain to break down,—would have led to hopeless confusion, and discredit.[83]

The postal system—a human, physical system—ended up being malleable and functional in a time when technical infrastructure was not. It had a well-honed work practice that it used and improved on in creative ways. It was the post office that eventually delivered the news between Sarah and George when the telegraph wires became inoperable. The sociomaterial practices associated with mail forwarding were integral to the functioning of public information infrastructures.

Registration and Information Bureaus

The fire that followed the earthquake destroyed the residences of about half the population of San Francisco—approximately two hundred thousand people. Those people dispersed throughout the Bay Area, and subsidized train service from Southern Pacific facilitated people traveling to other areas around the country. The disaster meant that many families were separated as they fled the fire to the west side of San Francisco, to the south to Daly City, to the east across the Bay to Oakland and Berkeley, and to the north to Marin County. A tragic consequence of hasty goodbyes was that people did not know where their loved ones had gone. Furthermore, employees and employers often had no notion of each other's whereabouts if a business was burned. An individual's entire social geography radically shifted along with their house. Below I focus on how people found each other in the days and weeks after the earthquake and fire using registration systems. Registration was a set of infrastructuring practices that recorded and organized the new addresses of people affected by the earthquake. The registration systems, as a public information infrastructure, were intended to centralize knowledge about people's whereabouts after the earthquake. Displaced earthquake publics were constituted in public information infrastructures through the process of registration, which (ideally) positively shaped people's experience of the earthquake by helping them find their friends, family, and colleagues.

As described in the preceding sections, the post office and telegraph were inundated with people trying to get in contact with each other. Newspaper columns were ideal places to advertise one's whereabouts and needs because of the centrality of the newspapers to the earthquake publics. The publishers of daily newspapers had built iconic buildings in downtown San Francisco before the earthquake to house their publications, and they didn't intend for their presence to disappear, even though their monuments had been badly shaken and burned.[84] All the printing facilities for the daily newspapers in San Francisco were destroyed, though. The *News*, whose headquarters was not in downtown, was the only San Francisco newspaper to print an edition during the short time period after the seismic activity and before fire struck. With a public hungry for news, the *Call*, *Chronicle*, and *Examiner*, three of San Francisco's four major newspapers, shared the *Oakland Tribune*'s presses for one day following the earthquake. After that one issue, the four newspapers (including the *Bulletin*) found temporary homes on different Oakland presses.[85] Impressively, the organization of the newspaper businesses was such that even with entirely different facilities, the daily San Francisco news companies were able to produce their own newspapers just three days after the earthquake.

Despite all the daily San Francisco newspapers having to relocate their printing facilities, the newspapers remained the best way to contact loved ones or promote civic messages. Moreover, the earthquake and fire made it clear that having functioning facilities and printing apparatus in San Francisco was not necessary for newspapers to retain their power. The newspapers, central for reaching many people, were a useful place to broadcast messages.[86] Because the newspapers were circulated widely, they were a recognizable place to publish or find traces of people's movements. On April 21, three days after the earthquake, when the fire was still burning, city newspapers started printing lists giving the whereabouts of people and businesses, or seeking information about others.

These lists included "where refugees have found shelter" along with the names of missing persons, patients in hospitals, and the deceased. Beyond the advertisements that people put in the different newspapers, many English-language newspapers also set up "registration bureaus," organized "to relieve the terrible mental strain of those parted from relatives and friends."[87] For example, the *Call* set up a registration bureau where refugees were located. It started registering refugees at the Baker Street entrance to the Panhandle and printing the messages in the April 21 issue.[88] Across the Bay, the *Chronicle* published lists of refugees who registered at its Oakland office.[89] In one of her letters to George, Sarah explained that she might need to go to *all* the registration bureaus to look for a friend of hers; refugees might be registered at just one bureau or several. In addition to newspaper advertisements and registration bureaus, fraternal organizations, relief committees, and the police all set up registry bureaus around San Francisco and the surrounding areas to which people had fled. The idea was that if people wanted to be found, they could go to the registration bureaus, where they would give their old and new addresses. People then could inquire as to the whereabouts of their families, friends, or work contacts using their names and old addresses.

The existence and purpose of these bureaus was publicized far away; the *Washington Times* noted that the "bureau of registration is bringing many families together."[90] The task was Herculean. Newspapers begged that the over two hundred thousand people displaced by the earthquake and subsequent fire let the registration bureaus know of their whereabouts in the "briefest possible manner" as well as the "concisest fashion possible," such that "lists may be prepared and published in the morning papers."[91] The *New York Tribune* called the task "illimitable and almost impossible," saying that was "occupying hundreds of persons [in one night]."[92] The *Los Angeles Herald* called the registration "an important piece of work" for which in Oakland, "forty-five clerks engaged in this work alone."[93] The early success of the registration bureau was quite obviously mixed, with one paper reporting, "Even with the registration bureaus many persons are still unable to find relatives or friends," yet the bureaus maintained that "registration is the only systematic method of bringing together separated families."[94] As people scattered about the state in the days immediately following the earthquake, between April 21 and 23, the newspapers devoted columns to these kinds of personal advertisements. Furthermore, the newspapers pointed people to locations where they might register. In a single issue of the *San Francisco Bulletin*, on April 21, 1906, the newspaper listed the ways for people to locate each other: via registration bureaus hosted by a different institutions, the relief committee, the *Bulletin* itself, fraternal organizations, or if they had any connections outside San Francisco, their hometown newspaper.[95] The space dedicated to printing lists of "registered" people dwindled in the issues of the *Bulletin* on April 24 and 25, although there were still personal advertisements or listings of "missing friends" until mid-May.[96] Even several weeks after the earthquake, the newspapers still encouraged registration: "All persons looking for missing friends are requested to register with the bureau, giving the names and if possible, the former addresses of the persons sought. Those present in the city at the time of the disaster are also asked to register, in order that their present whereabouts may be known by those desiring to communicate with them."[97] Registration was part of the process of public information infrastructure that enabled people to account for one another.

Outside the Bay Area, refugees were encouraged by the newspapers to register to facilitate notification of well-being and glorify the newspaper. The *Los Angeles Herald*, like many of the Bay Area newspapers, set up a registration system to enable people to find refugees who had fled to Los Angeles. The *Herald* advertised, "Experienced newspaper men have been detailed to devote their entire time to the directory and the names and present addresses of thousands of those who were made homeless."[98] The next day, the *Herald* reported on the success of its enterprise, declaring that "hundreds of names of the San Francisco refugees and of those seeking news of friends and relatives known to have been in the devastated city or in the earthquake and the terrible scenes that followed were added to *The Herald's* directory yesterday."[99] The paper lauded itself, saying, "Many letters of thanks for aid extended have been received by the bureau."[100] The blunt self-promotion makes it clear that newspapers were not running the registration bureaus for purely humanitarian reasons. Furthermore, wealthy men such as *Examiner* owner William Randolph Hearst were able to set up their own relief stations.[101] It was an opportunity for newspapers to insert themselves into the personal lives of those affected by the earthquake.

The newspapers both set up their own registration bureaus and advertised other bureaus, including those run by nongovernmental fraternal and relief organizations, but it also seems that there was an "official" registration bureau, though how the various newspaper, fraternal organization, and official registration bureaus related to one another is unclear.[102] By the time the fire had stopped burning, satellite registration bureaus were set up in Oakland and San Francisco at the edge of the fire-damaged area to accommodate all the people trying to make their personal information available and locate others.[103] Newspapers advertised the locations of a "main" registration bureau in San Francisco and these satellite registration bureaus in the few days after the fire stopped.[104] Despite the fact that these "new" information systems in the form of registration bureaus appeared to help people get in touch with others, it was the already-established newspaper companies that supported them.The registration lists from satellite bureaus in San Francisco and Oakland were to be forwarded to the main bureau in the mayor's office. One newspaper described the system thus: "Branch registries have been established at all points to which people have fled for refuge and their lists will be sent to the San Francisco main office as promptly as possible, and from these lists a general list will be prepared which will be printed in the papers as soon as possible."[105] The idea was that people would register by giving their old name and address along with their new address. According to one article, the main registration bureau at one point had thirty-five thousand names in it and got three thousand new names every day from the satellite bureaus; while this number is obviously a rough estimate, it does point to the potential volume.[106] The documentation about the registration bureau indicates that the bureaucratic office technology worked as follows: "Each name is written on a card at the central bureau with the two addresses [pre- and postearthquake] and these cards are filed alphabetically in cabinets so that it is possible to find a name and address with no loss of time or energy."[107]

Despite being widely advertised in newspapers, the official bureaus may not have gained much traction. When General Adolphus Greely, the commander in charge of the Pacific Division of the US Army in San Francisco, reached San Francisco a few days after the earthquake, he made an initial report to the secretary of war in Washington, DC, about the state of the city. Greely was asked for strategies to deal with the large volume of requests that the US Department of War was getting from people trying to locate others. He recommended that people reach out to the large daily newspapers and noted that there was also a registration bureau in Oakland.[108]

As public information infrastructure, the English-language newspapers did not equally serve all San Franciscans. The advertisements printed in newspapers were largely focused on those who found themselves in a residence with an address, knew people with an address they could use for communication, or had a permanent residence before the earthquake.

Many working-class San Franciscans lived in boardinghouses before the earthquake and refugee camps after it, and thus may not have had stable pre- or postearthquake addresses to point to.

Moreover, groups experiencing racial discrimination within San Francisco were rarely able to access the dominant public information infrastructures, even before the earthquake.[109] After the earthquake, however, these groups often found their own work-arounds. Many San Francisco businesses owned by English speakers used the English newspapers to notify people of new business locations after the earthquake; Chinese-run businesses used Chinese-language newspapers for the same purpose. Handwritten issues of *Chung Sai Yat Po* also advertised a separate system by which Chinese people could locate their friends and contacts: bulletin boards for people to leave messages for friends and relatives.[110] This speaks to the relationality and multiplicity of public information infrastructures. While the English newspapers may have been an ideal place for English speakers to broadcast notifications, it would not have been central to the Chinese-speaking public. Because not everyone experiences infrastructure the same way, public information infrastructures can be multiple. This contributes to narratives by other researchers suggesting that the disaster did not strip away lines of class, as many believed in the days after the earthquake, but rather exacerbated and reified them.[111] The stories of the infrastructures serving the Chinese citizenry illustrate the ways that preexisting alternative public information infrastructures persisted and adjusted to new circumstances. It also demonstrates the racialized manner in which public information infrastructures articulate in the information order.[112]

Design of the Red Cross Registration System

Once the first week after the earthquake had passed, notices about registering became less about the registration of people for the purposes of locating one another, and more about how to register for the purpose of receiving relief or unemployment benefits. This second wave of registration bureaus was not ad hoc or following the well-understood conventions of newspapers personal advertisements; it was designed. In addition, while the first set of registration bureaus focused on connecting people, the next set concentrated on making people legible in order to distribute resources.[113] While the registration bureaus for locating people were run by institutions that were highly visible to earthquake publics (such as newspaper companies), institutions already involved in bringing people together (such as fraternal organizations), or institutions seen as officially trying to help (such as the registration bureau at the police headquarters or relief offices), the registration for food was run by the Red Cross, an organization not involved in the day-to-day lives of most aid recipients before the earthquake.

The registration bureaus for notifying and locating others relied on individuals coming to the bureaus, whereas the registration process for distributing food had volunteers canvassing refugee camps. Starting as early as April 25, 1906, San Francisco newspapers reported on the development of a food distribution system run by the Red Cross Society. Some of the ideas behind the organization of the food distribution and corresponding registration system were to locate distribution centers in the appropriate locations, avoid having any substation serve more than a thousand people (in order to avoid long lines), and ensure that the stations had enough food (at least this was the system initially advertised in the newspapers).[114] Another stated purpose of the Red Cross registration system was to "furnish general statistical information of the progress of relief."[115] A last goal of this vast registration system was to reduce what the Red Cross perceived as "waste" in the system—that is, people receiving more than their fair share.[116] This was accomplished by doing almost the opposite of what was initially discussed. It centralized the relief effort so that there were fewer distribution stations, resulting in more people waiting in line. Even as people waited in line, though, they were counted.

The first Red Cross registration was designed by Edward Devine, who was sent to represent the Red Cross on the San Francisco Relief Committee, and Carl Copping Plehn, a professor of commerce at the University of California at Berkeley. As described by Red Cross historian Jones, Devine was working from precedent; he had analyzed the Chicago relief program following the fire of 1871.[117] Devine, an academic and secretary of the New York Charity Organization Society, was ready to apply the lessons in "scientific charity" and paternalistic antipauperization he had learned from Chicago to the situation in San Francisco.[118] After the Chicago fire, relief workers devised a system whereby the city was divided up and cases registered. Centralizing registration had been an effective way to systematize relief work and identify those who needed aid versus the impostors; this system would be imitated in San Francisco.[119]

Plehn had experience working on the census in California in 1900 as the supervisor of the first district, which included San Francisco, Alameda, and Contra Costa Counties. The 1890 census in San Francisco was "universally condemned" as "instances of careless work were so numerous that they excited no comment except disgust."[120] Newspapers cited several challenges in population enumeration: counting homeless people was difficult when the census was done by going door to door; people living in boardinghouses and other temporary residences were ignored; migrant workers were out working during the summer months; and many wealthy people were out of town for work and leisure during the summer months when the census was conducted. After the debacle of the 1890 census, Plehn was determined that his enumerators would find every single San Francisco inhabitant and each would be counted. Plehn described in a newspaper article how his census enumerators doggedly tracked down every single San Franciscan, calling in the police when they thought it would help their case.[121]

Devine, an analyst of charity work, and Plehn, an experienced supervisor of the census in San Francisco, collaborated to run the registration department in the Red Cross. Yet in his later writings, Devine gave Plehn much of the credit for design implementation.[122] Their goal was to design

Figure 3.9
Close inspection of the signage in an image from the refugee camps at Golden Gate Park shows that at least one information bureau was colocated with Red Cross tents and opportunities for free transportation on the Southern Pacific. Image: Public health and emergency relief efforts in San Francisco (graphic), BANC PIC 1994.022-A-B v.1:02. Courtesy of the Bancroft Library, University of California at Berkeley.

"the complete system of registration now organized in connection with the issuing of supplies." They describe the registration system in a five-page typed document called "The Plan for Registration," which was later reproduced in briefer form in the *Red Cross Bulletin*.[123] The document begins with a straightforward explanation of the goal of registration: "The purpose: In order to unify the meods [*sic*] of relief, to regulate the issue of food, to keep a record of the work done at the various Relief Stations, and to facilitate the centralization of the relief work, it is necessary to enroll all applicants in a general register." The idea was to divide the city into different districts, each with identical procedures managed by a "Central Registration Bureau of the National Red Cross."[124] Each family would get a registration card, which would be filled out "by an executive official at the Station" or "a canvasser of the Associated Charities."[125] The information on the registration card would have to be verified with a "visit to the place where the applicant" was living to prevent duplication of registration.[126] The final registration guide appeared in the *Red Cross Bulletin*: "More than nine out of every ten of the applicants will be self-supporting in a few weeks. The few lazy imposters [*sic*] will be speedily detected and dealt with separately. Assume every one to be entitled to relief until clearly proven unworthy."[127] This reminder appeared many times in instructions to relief workers and is indicative of an attitude that people should be able to pull themselves up by their bootstraps in short order. These ideas of the "worthy" and "unworthy" poor were baked into the system for accounting for people, shaping how the earthquake publics came to both understand and experience who was deserving of help. The agendas of elites such as Devine and Plehn influenced how the earthquake publics in need of aid were represented in the design of these registration systems.

Reception of the Red Cross Registration System

The Red Cross "system" of registration was relatively straightforward: divide the city into districts, centralize relief in those districts, and canvas them.[128] Before this organization system, other relief work around San Francisco had been difficult for the city elites to manage from the top down.[129] Now, however, they were able to manage it efficiently. The centralized distribution system made sure that every person was only receiving one portion of aid.[130] On the surface, the registration system may have seemed laudable to the charity workers, but they found implementation difficult. One relief worker wrote, "It can hardly be imagined what a colossal task it has been to merely register the persons receiving relief."[131] Unlike the census, where the work would take place over months with a trained workforce in a mapped area, here the group performing the aid registration was frequently working on unfamiliar terrain. The registration materials were supposed to be part of what was used by the relief workers responsible for determining how much aid, beyond food aid, people received.[132] But these paternalistic notions of charity were not always well received by aid recipients.

The registration system of subdividing, centralizing, and registering by the Red Cross inflamed some San Franciscans.[133] The centralization, General Greely argued, "enabled each applicant to get food and supplies at a sub-station without flocking promiscuously either to headquarters or to the central store-houses."[134] However, by centralizing the relief, ostensibly to prevent fraud and give everyone access to similar service, the system of registration also forced people to spend long hours in line. Physician Margaret Mahoney wrote in a frequently quoted pamphlet a few months after the earthquake, "The cry of fraud; the cry of famine; the hours of standing in line; the endless circuit from one relief station to another in search of the necessaries of life, which were often never obtained; all these were systematic means of conserving the supplies."[135] It was not Mahoney's imagination that the system was made to reduce the number of applicants, forcing them into endless lines. This experience seems to have been by design. "Before the registration it had been the custom for the refugee families to send as many children to the bread lines as possible, thus securing excessive relief."[136] The registration did what it was intended to do: decrease the amount of relief distributed by making the process uncomfortable.[137] While Devine and Plehn enacted familiar information practices with the registration system, for San Franciscans it was unfamiliar, and for some, unfair and unwanted.

Historians Davies and Jones documented the resistance to the progressive character of these charitable systems via the actions of the group United Refugees, incorporated in July 1906, which argued that relief should be divided equally among citizens, not allocated based on the judgment of the Red Cross elites or what people had before the earthquake.[138] By the time of the earthquake, San Francisco had built "the strongest labor organization in America," setting the stage for the radical, quasi-socialist aid distribution approaches of the United Refugees.[139] The refugees' complaints about the aid distribution process, along with the general suspicions concerning the fiduciary trustworthiness of San Francisco politicians, nonetheless did succeed in attracting national attention to their cause.[140] Two investigations concluded that there was no wrongdoing by the relief committee, and "modern methods prevented the misappropriation of relief funds."[141] Although the protests were mostly unsuccessful, it is clear that everyone did not share the progressive vision implemented by the relief committee.

While the registration system was met with contempt from some, Red Cross workers' enthusiasm for it can hardly be understated even though it was challenging to implement. The system was suggested as a potential model for the future and kept on file at National Red Cross headquarters.[142] General Greely implied in later reports that the San Francisco model became immediately influential across the country as a best practice for handling disaster relief.[143]

Counting the Dead

Public information infrastructure as a set of sociotechnical processes involved in producing and circulating knowledge about individual people's welfare impacted the earthquake publics' experience of the event in the aftermath. Different and competing information agendas ultimately shaped the information order, and by extension, the earthquake publics' knowledge and experiences of the event. Nowhere was the information order's effect on event epistemology more obvious than in the process of counting the dead.

Many seem to recognize that from the moment the fire had gone out and people began to look for loved ones, figuring out who had died would be a difficult task. Initially, registration bureaus advertised that they sought names of the deceased.[144] The newspapers were quick to print names under bold headlines as they were released to the public. The day of the earthquake, the *News*, the only San Francisco newspaper to print an edition after the seismic activity and before fire struck, printed a list of the known dead. The *Chronicle* declared, "Facts will never be known as many lie in unnamed graves."[145] Furthermore, bodies started decomposing as they were dug out from the ruins, making it necessary to bury them quickly.[146] On April 27, a request by Greely was printed widely that seemed to indicate that he endeavored to make a systematic count of the dead: "All hospitals, doctors, coroners, and others concerned" were to report information about the deceased and injured to a military officer at Fort Mason "for the purpose of making a correct list ... so that it may be published [and] broadcast to satisfy the inquiries of thousands all over the United States."[147] On April 28, newspapers all over the country printed something like "Greely's Death Roll."[148] While newspapers pronounced in bold headlines the "List of Dead Is Increasing," they also seemed to cap their expectations: "The likelihood of the total number of fatalities exceeding 500 diminishes."[149] The official report by Greely said, "304 known; 194 unknown (largely bodies recovered from the ruins in the burned district)" died in San Francisco alone. Greely reported 166 dead outside San Francisco.[150]

The Subcommittee on Statistics, part of the Committee of Forty on Reconstruction, convened in the months following the earthquake and endeavored to count the dead. The *Report [of] the Sub-Committee on Statistics to the Chairman and Committee on Reconstruction* contains a letter signed by 6 people that says there were 322 "killed outright and accounted for at the Coroner's office," and 352 "reported missing and not accounted for," for a total of 674 dead in San Francisco.[151] These numbers were surprisingly low, even to those doing the counting who remarked that the number who died was "comparatively slight" owing "to the spirit of the people and to the directing forces brought to bear upon the catastrophe."[152] The stated desire to carefully register and count refugees needing aid makes the inaccurate counting of the dead all the more intriguing.

Sixty years later, Gladys Hansen, an archivist in San Francisco, argued that the number of dead was too low for several reasons.[153] Some accounts described the collapse of large hotels and buildings occupied by the working-class poor south of Market Street, an area particularly vulnerable to liquefaction. These buildings would have sunk into the earth, trapping people inside before burning. Moreover, buildings burned at a temperature that left little in the way of identifiable remains. Other anecdotal evidence suggested that dead people's bodies placed in a temporary morgue were incinerated when the fire swept through, leaving only time for the living to escape.[154] Hansen reasoned that one way to get a more accurate count of the dead would be to document the names of the deceased using a variety of methods.

Hansen's multidecade exploration of the archives divulges what records were available. Her initial studies examined newspaper reports of missing people against city directories.[155] She added a number of other public records, including death certificates, coroners' reports, Edwards Abstracts, McEnerney cases, inheritance tax records, city and county records of orphans, San Francisco National Cemetery records, and Presidio, Harbor, Park, and Central Emergency Hospital records.[156] The private records of the mortuary J. C. O'Connor and Company were also donated. Along with her associates, Hansen found that there were at least 992 nameable deceased and 75 unknowns.[157] Authors extrapolating from Hansen's work suggest that the death toll could be in the thousands, considering her count included mostly named deceased reported in newspapers and other official sources.

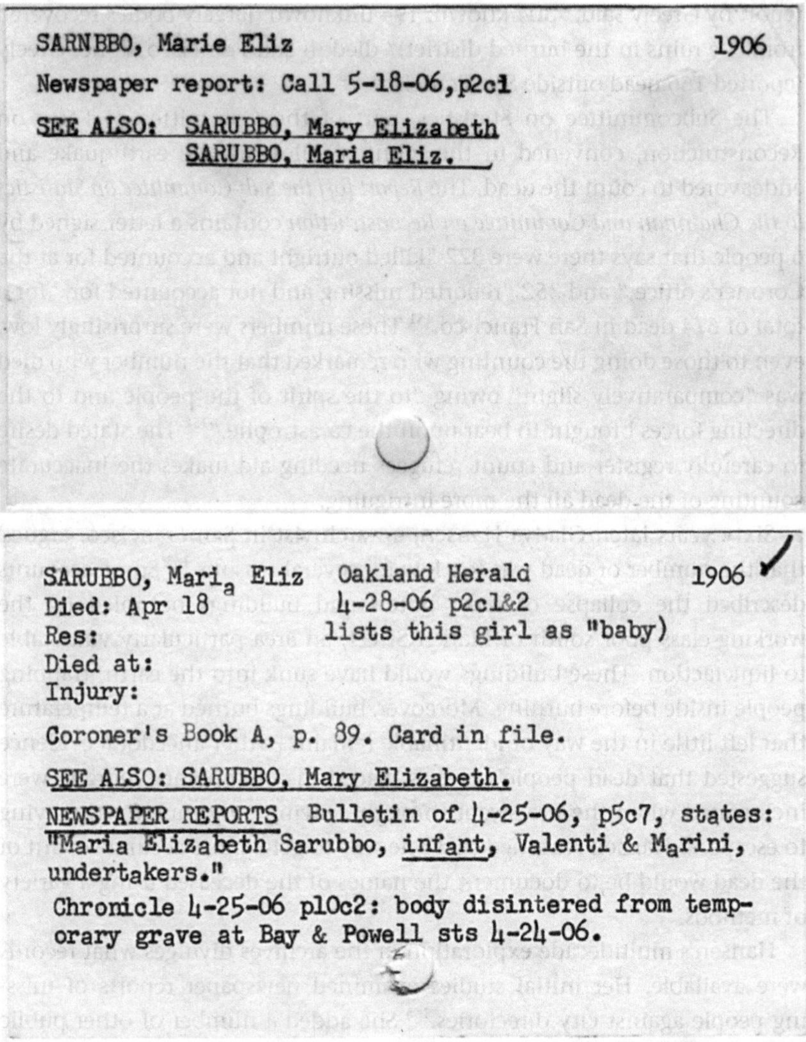

SARNBBO, Marie Eliz   1906
Newspaper report: Call 5-18-06,p2c1

SEE ALSO: SARUBBO, Mary Elizabeth
           SARUBBO, Maria Eliz.

SARUBBO, Mari$_a$ Eliz   Oakland Herald   1906
Died: Apr 18   4-28-06 p2c1&2
Res: lists this girl as "baby)
Died at:
Injury:

Coroner's Book A, p. 89. Card in file.

SEE ALSO: SARUBBO, Mary Elizabeth.

NEWSPAPER REPORTS Bulletin of 4-25-06, p5c7, states:
"Maria Elizabeth Sarubbo, infant, Valenti & Marini,
undertakers."
Chronicle 4-25-06 p10c2: body disintered from temporary grave at Bay & Powell sts 4-24-06.

Figure 3.10
Archivist Gladys Hansen had three cataloging cards about Maria or Mary Elizabeth
Sarubbo, who was apparently an infant at the time of her death. Hansen's records
from the San Francisco Public Library about the Sarubbo baby show the multiplicity
of records in use in establishing death as well as the traditional library cataloging
practices Hansen used to account for the dead. Image: San Francisco History Center,
San Francisco Public Library.

SARUBBO, Mary Elizabeth (infant)*   1906 ✓✓
Died: ᴬpr. 18

Dept Pub. Health records - Coroner's Book A, p.89

* Bulletin 4-25-06 p5c7: "Maria Elizabeth Sarubbo,
  infant. Valente & Marini undertakers"
See: SARUBBO, Maria Eliz. Coroner's Book A p 167
Chronicle 4-25-06 p10c2 states: Body disinterred from
temporary grave Bay & Powell sts 4-24-06

Figure 3.10 (continued)

RUGGLES, John W.   1906

John W. Ruggles... committed suicide on the morning of
May 1st by shooting himself through the head at his
home near Mount View. Mr. Ruggles was injured by the
fall of a heavy beam at the time of the earthquake and
it is thought it unsettled his mind. (Retail Grocers'
Advocate, May 4, 1906, p.10.2.
S.F.Chronicle, May 2, 1906 p.10c.2: Mount View was in
San Rafael. Ruggles was secretary of the Dodge-
Sweeney Co., 114-16 Market St. & 11-13 California St,
San Francisco. 1905 city directory gives home as
Ross, Marin County.
Oakland Tribune 5-2-06 p.5c.3 also carries this story.
Filed also under "Suicides"

Figure 3.11
Even with Hansen's expert archival practice, the problem of what deaths resulted
from the earthquake was not always clear. Card from Hansen's catalog of the deceased. Image: San Francisco History Center, San Francisco Public Library.

Even with the newspapers releasing lists of the dead, and military requests for lists, the original death count was not accurate. Yet the registration and information bureaus would have had ample data from which to start, as they were initially advertising for people to report names of the dead.[158] Hansen and others charged that the Southern Pacific Railroad and the press made a deliberate effort to deny that the earthquake was destructive and promote the idea that damage was caused by fire.[159] The earthquake was thought to be bad for the image of San Francisco.[160] Blaming the fire for the damage served two purposes. First, many Californians had fire but not earthquake insurance, and thus they could be compensated if their property was thought to have been destroyed by fire.[161] Second, unlike seismic safety, fire prevention was being addressed in many US cities. In other words, fire was a familiar horror that many US experts had studied after recent conflagrations.[162]

Newspapers and other "boosters" denied that seismic hazards were serious, and urged the rapid rebuilding of San Francisco.[163] One newspaper published a letter appearing in a California newspaper suggesting that people write cheery letters to correct sensational stories appearing in East Coast newspapers and assure readers that San Francisco was rebounding.[164] Kevin Rozario argues that the narratives of renewal seen after the earthquake were encouraged because they fit into a Progressive vision of society: the disaster meant a clean slate for improvement.[165]

The earthquake affects were also underreported because, as Hansen and Condon contend, many of those who died lived in a district of San Francisco built on human-constructed land populated by poor immigrants.[166] Hansen and Quinn also point out that in their effort, they could identify the names of only twenty-two Chinese and six Japanese—extraordinarily low numbers considering that they lived in densely populated areas where buildings were in disrepair.[167] Many of San Francisco's poorest and racially marginalized were those living in the worst buildings and not in the forefront of political consciousness before the earthquake, nor were they afterward. These people were unknown to the institutions that might have identified them before the earthquake and hence disappeared forever. Modern efforts to count the dead still leave out those who were marginalized.

Whether or not there was a deliberate attempt to downplay the effects of the earthquake, there was certainly reason to underestimate the number of deceased. This seems rational from the perspective of redevelopment interests in San Francisco, but seems deceitful when weighed against ideals insisting that the dead should be recognized and accurately accounted for. While the disaster did mean that many reports and accounts of the disaster were produced, a system of finding others was developed, and a system of registering people receiving benefits was designed and executed—all documentary trappings of a reflexive modern society—the dead were not systematically recognized. Drew Gilpin Faust makes the argument that the naming and counting of dead following the Civil War was a complex political effort undertaken by a multitude of organizations.[168] Thus, efforts to count the dead were not unheard of after massive destruction, but there had to have been either personal or political will involved. Counting and naming the dead was simply not an activity that the military, local government, or any of the relief organizations found politically advantageous. The information order that produced this event epistemology has made it nearly impossible to count all of the dead, even today.

Conclusion

As George attempted to find out about the well-being of Sarah, and Sarah tried to notify George as to her personal status via telegrams, the few operating telegraph lines were overwhelmed. The information needs of the earthquake publics—to learn about the fates of the others impacted by the earthquake—were substantiated in the public information infrastructures produced by media institutions, relief groups, government organizations, and the earthquake publics.

The dispersal of half the city's population, sudden need to contact loved ones, and destruction of the city's physical newspaper and telegraph infrastructure were hugely disruptive, yet the manner in which public information infrastructures were reconstituted reflected the social, political, and economic practices of the participants, and was articulated in the information order. The commercial telegraph business worked quickly to get cables running again, but so many had filed for telegrams to be sent that the telegrams had to be sent by mail. The post office reconstituted mail service quickly and delivered many of the telegrams. While physical infrastructure was burned and destroyed, the way that people organized and worked was not, and that helped the information infrastructure recover. "Old" (pre-earthquake) public information infrastructures were broken or insufficient for helping people get in touch. "New" information systems in the form of registration bureaus appeared to help people get in touch with others. Yet it was old newspaper companies that advertised them. Relief organizations registered people to track their use of resources and were met with resistance by some San Franciscans. While displaced persons were accounted for by relief organizations, all the dead were not counted until decades after the earthquake.

Stories about the operability of the telegraph, post office, and registration bureaus as well as the enumeration of the deceased implicitly suggest a narrative of institutional continuity. The postearthquake information infrastructure for accounting for people often reflected predisaster methods and systems for gathering and sharing information. Deep-pocketed institutions (such as the newspaper companies) were able to recover quickly, and because of the pressing need for news after a disaster, this reinforced their centrality in the public information infrastructure.

Within the information order, however, Bay Area residents had radically different experiences. The information-lacking earthquake publics constituted by and embodied in public information infrastructures were not inclusive. Elite diplomats had the governor tracking down loved ones, while the poorest earthquake publics were subject to humiliating record-keeping practices or simply ignored. Alternative public information infrastructures supported non-English-speaking San Franciscans, but were not dominant in the information order.

4 The 1989 Loma Prieta Earthquake: "Public Information" for Alternative Earthquake Publics

When the Loma Prieta earthquake struck on October 17, 1989, San Jose State University students in an introductory Mass Media and Culture class were working on individual "media consumption" diaries. The students were asked to self-report when they made use of media over a five-day period and reflect on their practices. One student reported in her diary that she turned to the media almost immediately after the earthquake: "After I was sure I was OK and so were my neighbors, I ran upstairs to find our battery operated radio. No one had one in our apartment complex but me!" This student continued to be enthusiastically interested in the earthquake, writing, "I can't wait to hear what's up. We have no electricity. But thank God we do have a battery operated radio!"[1]

As one of the only people in her apartment complex to have a battery-operated radio, the student said she updated her neighbors as the news rolled in. She headed to bed emotionally drained, and when electricity was restored the next day, turned on the television, anxious to "see the effects with [her] own eyes." She was shocked to see the earthquake on television: "Jesus! Quake looks like it did a lot of damage."[2]

Reports on the most severe (and exaggerated) damage quickly became a challenge as phone calls from family and friends start pouring in. One student wrote, "Phone calls from Parents in L.A., Mother-in-law in N.Y. Friends from Hawaii & Modesto. I tried to calm everyone but it is too late. The media really scared my family. My husband and I are scared too." Eventually, two days after the earthquake, on October 19, this student began to feel "burned-out" after "listen[ing] to news on TV for so many hours." She bought a copy of the *San Francisco Chronicle* the next day, expecting more accurate and measured coverage than what had appeared on television, but was disappointed. "Bloody pictures were everywhere. It looks like the whole city was up in flames."[3] Just as I have explored in previous chapters, the information order of the day greatly influenced the earthquake public's experience of the earthquake.

When the magnitude 6.9 Loma Prieta earthquake occurred approximately sixty miles south of San Francisco, it caused an estimated $10 billion in damage and the deaths of sixty-three people, including forty-three from the collapse of a highway in Oakland.[4] At the time, Loma Prieta was considered the costliest natural disaster in US history. The government's role in disaster response enlarged throughout the Cold War, and by 1989, the government imagined its work as a producer of authoritative public information about the earthquake.[5] In contrast, in 1868 and 1906, no explicit plans existed for the government's postearthquake response. With the growth of the state's role in crisis response came an expanded place in the postdisaster information order, and by 1989, the government imagined its work as a producer of authoritative "public information" about the earthquake. Furthermore, technologies such as radio and television enabled the mainstream commercial media's ability to shape knowledge of the 1989 earthquake. In places where electricity was available, the media instantly reached people and told them about what had happened.

Loma Prieta is a turning point from the previous Bay Area earthquakes described in this book because of *who* claimed the authority to dominate earthquake knowledge. In 1868 and 1906 in San Francisco, the newspaper owners and business elites, often themselves members of the local government, attempted to propagate their own narratives of what happened after the earthquakes.[6] The 1989 earthquake prompted the enactment of a large, planned response from FEMA at the national level, the California State Office of Emergency Services at the state level, and local municipal emergency response organizations. Members of these organizations included assorted professional civilian emergency responders—jobs that didn't exist in 1906 outside the police and fire departments. I argue that the 1989 professional responders envisioned shaping the event epistemology after a disaster through the production and circulation of public information.

I use government disaster plans, described in the first section below, and the postdisaster reports issued by various government groups and academics, funded by the US government, as guides to understanding the government's earthquake response.[7] The government's preparation in 1989 is largely reflected in the state disaster plans, which in turn, reflect the state's assumptions of what the people of California would do after a disaster. These disaster plans describe how emergency responders should produce "public information." The vision of public information in these disaster response plans was a product of the civil defense era plans and thinking, emphasizing the public as receivers and the media as transmitters of the government's message.[8]

This vision, however, did not correspond with what actually happened after Loma Prieta. The plans did not take into account that television, radio, and newspaper companies would suffer damages from the earthquake and have difficulty operating. Furthermore, the plans did not anticipate that the media would follow their own story, focusing on the most damaged areas of San Francisco (the Marina) and Oakland (the collapsed highway, known as the Nimitz Freeway or I-880 Cypress Street Viaduct). Government disaster plans envisioned that they, the official disaster responders, would be the ones generating authoritative accounts for all the people affected, and the media would provide an unproblematic conduit between the public and government officials—yet this did not happen. The government didn't take into account the agency of the media and public as well as its own limitations after an earthquake. This chapter shows that immediately after the earthquake, the government officials who were supposed to supply public information did not have the reports they expected from the field, and the officials themselves often relied on the media to get a sense of what happened. Still, the media frequently cited the professional disaster responders. The state and media reinforced the centrality of each other's informational role in disaster response: the media sought out the disaster responders as official sources, and the disaster responders sought out the media to put forth emergency public information. In other words, the government's plans assumed that the government would have more control over people's knowledge of the earthquake than it did.

The plans also did not deal with the multiplicity of earthquake publics and the alternative public information infrastructures that would shape different Loma Prieta event epistemologies. These overlapping public information infrastructures constituted the information order and made certain types of knowledge possible while limiting others. This is reminiscent of the issues that arose with the multiplicity of publics that underpinned C. A. Bayly's original conception of an information order when analyzing the rebellion against the British in 1857 colonial India.[9] Understanding the dynamics among multiple publics producing multiple public information infrastructures is key to understanding the 1989 information order. In Bayly's work, he analyzes both Indian production and circulation of information, and British surveillance practices undertaken for the purpose of exercising power and control. Historians had argued that the British focus on using statistics obscured many of the Indian people's actions, such that when the Indians rebelled, the British were surprised because of British inattention to the Indian informal information networks. But in fact, Bayly notes that the rebellion also relied on the methods of document construction and circulation introduced by the British—the post, lithographed newspapers, and the telegraph. It was both the focus on statistical methods and British inability to see changing practices among the Indian citizenry that obscured their ability to make sense of what was going on. Using the terms of this book, in Bayly's colonial India, there were several public information infrastructures supporting different publics. The dominant public information infrastructure ignored the practices of alternative publics. As this chapter shows, in 1989, the different earthquake response plans reflected blindness to the information practices of marginalized groups.

After the 1989 Loma Prieta earthquake, people quickly discovered that the government disaster response plans were not realistic about who might be most adversely affected; in many cases, the disaster response plans or disaster responders had not considered the needs of marginalized groups such as non-English speakers and the poor in the public information produced immediately after the earthquake, or in the recovery services made available. In the 1906 chapter, I briefly discussed Chinese-language newspapers and how the transient would have been left out of many of the different schemes for accounting for people. Picking up from the 1906 chapter, the second half of this chapter further develops the concept of alternative public information infrastructures, using the case of how the Spanish-language media provided translations of government information and focused on stories about how people in their community fared after the earthquake. The earthquake public brought into being through this alternative public information infrastructure created a different event epistemology that profoundly shaped how this community experienced the earthquake.

Establishing Informational Authority

The Loma Prieta earthquake put a well-developed government disaster response apparatus into action. These disaster plans described the state's information infrastructure, particularly how the professional disaster responders working for various government agencies would produce and disseminate public information. In the vision of the disaster plans, this public information would be authoritative, meaning that it would assert the "correct" version of the story. Yet "the government" was not a monolith; the disaster responders included organizations dedicated to disaster response, such as FEMA at the federal level, and California state agencies, such as the Office of Emergency Services. In addition to these agencies, government employees from city and county management, police, public works, and fire departments activated disaster plans and served as earthquake response organizations. Professional disaster responders in organizations dedicated to responding to major disasters, such as FEMA and the Office of Emergency Services, worked on and at times used these disaster response plans.

All these disaster response plans envisioned that the government would play a central role in creating authoritative public information. In these plans, the visions of public information were even sometimes explicitly given.[10] San Francisco's 1988 *Multihazard Functional Plan* explained that "emergency public information" is "information disseminated to the public by official sources during an emergency, using broadcast and print media."[11] Though plans were activated and implemented to varying degrees, at all levels of government the disaster response plans in place before the Loma Prieta earthquake placed the government at the forefront of producing public information.

One of the key plans in place at the time of the Loma Prieta earthquake was the 1987 *Federal Response to a Catastrophic Earthquake*, published by FEMA, and designed to be used if state and local disaster response capabilities were overwhelmed and federal intervention was required. For the most part, within "the Plan" (as it called itself) there are two areas explicitly concerned with informing the public.[12] The first is through the work of FEMA "Public Affairs Officers," who reported to the field coordinating officer, the person appointed to manage the disaster response and serve as the media spokesperson.[13] The second area of the Plan concerned with public information was the Joint Information Centers, an organizational structure specified in the Plan as ideally working to "ensure the coordinated, timely release of accurate information to the news media and the public ... and coordinate with State public information operations ... to provide assistance/information to the general public."[14] In other words, the entities described in the Plan (public affairs officers and Joint Information Centers) were the federal government's mechanisms for producing authoritative public information. In the aftermath of the 1989 earthquake, though, the Plan was only activated in a limited capacity by FEMA, which considered Loma Prieta a "major disaster," but not "catastrophic."[15]

California's *State Emergency Plan*, however, was put into full implementation. After Loma Prieta, Governor George Deukmejian declared a state emergency, activating the State Operations Center and Region II Emergency Operations Center in Pleasant Hill.[16] Ten counties were eventually considered part of the disaster area.[17] The guiding publication for the *State Emergency Plan* was the *Multihazard Functional Planning Guidance* document published by the State Office of Emergency Services in 1985, designed to guide disaster planning at "all levels of government" in the state.[18] According to this document, "the Jurisdiction Emergency Public Information (EPI) Organization" would "prescribe procedures for ... dissemination of accurate instructions and information to the public" as well as "response to media inquiries."[19] The information would be primarily gathered through the network of "Public Information Officers," roles similar to those of public affairs officials at the federal level, located in local and regional organizations throughout the state.[20] "Status boards" were supposed to facilitate communication between different response functions and the public information officer, or between the public information officer and the media.[21] Although the public information officers were supposed to respond to inquiries from the public, the only direct instructions and templates for communicating directly to the public (as opposed to through the media) are embodied in several sample radio scripts, which describe hypothetical earthquakes.[22] The California 1989 *State Emergency Plan*, which fills a binder approximately three inches thick, reflects much of the *Multihazard Functional Planning Guidance* in the area of public information.[23] According to the *State Emergency Plan*, the Public Information Officer was supposed to "inform" the media and, "at the local level, to respond to inquiries from the public."[24]

The California *State Emergency Plan* indicated that local governments should be in charge of direct communication with the public. Cities within the disaster area also enacted their own disaster plans, though city disaster plans replicated much of the apparatus for creating public information prescribed at the state level. San Francisco's 1988 *Multihazard Functional Plan* followed the state-level *Multihazard Functional Planning Guidance*.[25] The city disaster plan had a specific definition of the media: a "means of providing information and instructions to the public, including radio, television, and newspapers."[26] The guidelines for emergency public information after a disaster included instructions to "release emergency instructions/information to the public as necessary through the media using Media Contact list," establish "media only" telephone numbers, and set up phone lines with prerecorded messages for the public.[27]

Thus, according to all these disaster response plans, different government organizations would be involved in producing an authoritative story about what happened. Public information officers in municipalities would ideally share information with other public information officers at places such as the Joint Information Centers to ensure that the released public information was correct and consistent. Then the media would disseminate this public information.

Only a few months before Loma Prieta occurred, government officials staged a massive earthquake response drill where they rehearsed these plans. Called RESPONSE-89, the $500,000 drill imagined that five thousand East Bay residents died after the Hayward Fault ruptured.[28] Hundreds of federal and state emergency response officials went to a building in Sacramento, and pretended to manage an emergency response in front of observing reporters.[29] The drill used the *Federal Response to a Catastrophic Earthquake* plan along with the California Office of Emergency Service's latest earthquake plan as well as scenarios developed by experts based on recent earthquakes in Mexico City and Armenia.[30] While the simulation did help with identifying weaknesses, the real earthquake on October 17, 1989, however, did not go according to plan or simulation.

Reporting the Disaster

The government emphasized the importance of the media postearthquake in disaster plans, but in the actual aftermath the media did not behave like a perfect conduit for the government to disseminate information to the public. First, the earthquake injured the media infrastructure, and people scrambled to find electric power to continue to print newspapers or do broadcasts. Second, the media focused coverage on the most damaged areas in San Francisco and Oakland, to the exclusion of some of the areas closest to the epicenter. In part, this was because so many media people were in San Francisco at the time of the earthquake, but also because the reporters themselves identified more with people in San Francisco. The media looked for their own story about what had happened after the earthquake; they did not simply follow what the government plans had intended them to say.

In general, though power failures and the lack of backup generators made it difficult for some radio, television, and newspaper companies to get back in action immediately after the earthquake, most used some combination of limited generator or battery power to muddle through. Yet not all the media companies had followed the advice of preparedness guides and as a result went without power for several hours. Most of the dailies outside San Francisco, such as the *Oakland Tribune* and *San Jose Mercury News*, had backup power and were able to print full editions the day after the earthquake. Newspapers that were printed near the epicenter, such as the *Watsonville Register-Pajaronian*, experienced damage requiring a number of work-arounds, but they still were able to deliver their editions.[31] While a few of the San Francisco newspapers accepted some of the offers from further-away newspapers to use their printing facilities, the two main San Francisco dailies, the *Chronicle* and *Examiner*, could not arrive at an arrangement to share emergency power generators and delivered their papers late. Despite this, "the lateness of delivery and the hunger for news about the earthquake had created an almost-insatiable demand for newspapers."[32] The *San Jose Mercury News* "printed more than 100,000 extra copies of the newspaper."[33] The great interest in news aside, however, the Bay Area newspapers were not as dominant as they had been in 1906—not just because the diversity of broadcast media now included radio and television, but also because other newspapers based outside the city served equally large and growing populations.

When representing the earthquake to a national audience, San Francisco remained central. The earthquake struck just before a game of the nationally televised Major League Baseball World Series at the San Francisco Giants' ballpark. The "live" aspect of the earthquake and fact that so many members of the media were at the Giants game made it seem to national viewers as if the earthquake was situated solely in San Francisco. The reports initially concentrated on San Francisco and then Oakland, and the earthquake was at first called the "San Francisco Earthquake."[34] It wasn't until an hour after the earthquake that television media outlets began to report that the epicenter was sixty miles south of San Francisco, and three hours after the earthquake, reports started to surface of how much damage had occurred in the city of Santa Cruz.[35]

The focus on San Francisco nevertheless persisted, even after the initial hour. A 1992 study by Conrad Smith of the three major national television news programs found that on the night of the earthquake, the television shots were overwhelmingly of San Francisco's Marina district. Smith suggests that the journalists' education and economic class made it easier for them to identify with the residents of the Marina district than with the poorer people from Oakland displaced by the earthquake or similarly marginalized people in other areas.[36] In a journalism symposium at the University of California at Berkeley a few months after the earthquake, members of the media gave a number of reasons for the attention on San Francisco, including "all of the earthquake mythology ... [associated] with San Francisco," the fact that San Francisco had more tourists and travelers than anywhere else in the Bay Area, the name San Francisco was recognizable worldwide, the World Series was happening in San Francisco at the time,

Figure 4.1
The collapsed section of the I-880 freeway in Oakland killed forty-two people. Image:
Two Mile Sandwich, taken on October 17, 1989, by Flickr user sanbeiji / Joe Lewis,
https://www.flickr.com/photos/sanbeiji/220647481. This photo is licensed under
the Creative Commons Attribution-Share Alike 3.0 Unported license. Some rights
reserved. https://creativecommons.org/licenses/by-sa/3.0.

and many non-California media outlets with representation in the Bay Area
had their branch offices located in San Francisco.[37]

In the weeks that followed, the national media was criticized for alarmist reporting of the earthquake. After the earthquake, the front page of a
number of newspapers, including the highly reputable *San Francisco Chronicle* and *San Jose Mercury News*, proclaimed that hundreds were dead. The
"273 dead," a figure widely circulated as confirmed on October 18, the day
after the earthquake, was in fact an estimate.[38] The inaccurate death counts
published by the newspapers and television stations were a source of anxiety and regret for the media, but also, as the following sections will show,
an illustration of the symbiotic relationship between these two entities.
Disaster response officials blamed the media for pressuring them to provide
figures about the number of dead and extent of the damage; the media, in
turn, blamed disaster officials for supplying incorrect figures.

The intense media focus had real consequences for earthquake survivors.
According to some city managers, media attention also resulted in pledges

San Francisco Chronicle

The Largest Daily Circulation in Northern California

125th Year No. 236   ★★★★★   WEDNESDAY, OCTOBER 18, 1989   415-777-1111   25 CENTS

HUNDREDS DEAD IN HUGE QUAKE

This three-story apartment house at Beach and Divisidero streets lay broken in the roadway as the big Marina District fire burned in the background
BY VINCENT MAGGIORA/THE CHRONICLE

The Experts' Advice on How to Cope

By Edward Epstein
Chronicle Staff Writer

As a stunned Bay Area comes back to life today, millions of residents will start trying to cope with the aftermath of the area's worst earthquake since 1906.

The advice from government and corporate leaders is that nonessential workers should stay home today from their jobs in the city. A spokesman for the State Office of Emergency Services said, "If you don't have to be here stay away."

Bill Newbrough added that roads and bridges into the city are expected to be clogged with as many as 200,000 East Bay residents coming to work in the city and unable to use the damaged Bay Bridge and Interstate 880.

San Francisco Mayor Art Agnos also said people should not go to work unless they absolutely must. If workers feel they must go to their jobs, Agnos suggested they call ahead to find

Page A 6 Col 5

The top level of the I-880 freeway at Cypress Street in Oakland fell onto the lower roadway
BY TOM LEVY/THE CHRONICLE

Oakland Freeway Collapses — Bay Bridge Section Fails

By Randy Shilts
and Susan Sward
Chronicle Staff Writers

A terrifying earthquake ripped through Northern California late yesterday afternoon, killing more than 200 people, injuring hundreds more, setting buildings ablaze and destroying sections of the Bay Bridge.

The earthquake was the strongest since the devastation of the great 1906 shock, measuring 7.0 on the Richter scale and shaking the state from Lake Tahoe to Los Angeles.

The temblor erupted from the treacherous San Andreas Fault and was centered in sparsely populated mountains 10 miles north of Santa Cruz.

Late last night, Lieutenant Governor Leo McCarthy estimated damage at "well over $1 billion" and predicted, "It will climb much higher in the light of day."

McCarthy said he had talked with White House Chief of Staff John Sununu who had activated 20 federal agencies to provide emergency relief.

About Today's Chronicle

This special edition of The Chronicle was produced despite a complete power outage at the newspaper's headquarters at 5th and Mission streets.

The earthquake brought the newspaper's computer and main printing facility to a halt.

The papers were printed at the Chronicle's Army Street printing plant and at its East Bay facility in Union City.

The news accounts were placed on Macintosh computer disks using an emergency generator.

"The Chronicle of October 18, 1989 is a newspaper produced with a heavy heart," said Executive Editor William German.

"The fact that it was produced at all was a mighty feat. This awful blow wiped out the tools of modern journalism. No magic of computers, no push-button presses, not even lights to see by.

"All that was left to us was the energy and the wit of our dedicated staff," he said.

The earthquake struck just as hundreds of thousands of people were leaving work for the afternoon commute, jamming freeway, filling mass transit systems and crowding city streets throughout the Bay Area.

Meanwhile, the eyes of the nation were already riveted on San Francisco, where the third game in the World Series was just

30 minutes away from the first pitch at Candlestick Park. The game was abruptly canceled.

The heaviest fatalities came about two miles south of the San Francisco-Oakland Bay Bridge when the upper deck of a mile-long section of two-tiered Interstate 880 collapsed onto the

Page A8 Col. 1

Figure 4.2

The front page of the *San Francisco Chronicle* on October 18, 1989, reported hundreds were dead. Image: *San Francisco Chronicle*/Polaris.

or donations.[39] The American Red Cross received an "unprecedented" $76 million, or three times the amount required to do its usual work of providing shelter, food, and basic medical services after an earthquake.[40] At the same time, though, the media attention and public donations created difficulties. In a 1991 report to the California Seismic Safety Commission, the Watsonville Office of the City Manager noted that "the public responded to media releases and literally dumped tons of clothing at the park and brought prepared foods, some of which spoiled without refrigeration. Conditions became a major concern to public health and safety officials."[41] Medical professionals also responded to appeals for assistance based on the alarmist reporting of the quake in the popular media.[42] Emergency workers voluntarily reported to duty in great numbers, and in general there were enough, or perhaps even too many, medical experts on hand to handle the relatively few injured patients.[43] The unrelenting national television spotlight on the iconic imagery of the Bay Bridge's collapse or Marina district's fire misrepresented the overall impact of the earthquake. Understanding what happened in areas that were not being covered by the media was difficult, and the lack of details about these areas further contributed to the focus on the Marina district and large number of fatalities as a result of the I-880 collapse in Oakland.[44] Media attention very much influenced both people's knowledge about the earthquake and their philanthropic responses. The event epistemology had real, material consequences for survivors.

Producing Public Information

The media performance after the earthquake did not conform to how the state envisioned it would in disaster response plans. Throughout the disaster response plans and drills of the 1980s, the state's imagined process of making public information cast the public in the role of receiver, the government as the authority with the official message, and the media as the willing transmitter of whatever the government envisioned. There were many flaws in this Cold War era notion of communication. First, the media was portrayed as an uncomplicated conduit that would not have its operations affected by the earthquake. Second, none of these plans accounted for the possibility that the network of professional disaster responders might be learning about the earthquake from television and radio—the presumed source of the same rumors that the officers were tasked with controlling.[45] Third, also absent was the idea that the public itself might have experiences that would be of interest to the professional disaster responders, who were supposed to only get information from other government officials. There

Figure 4.3
The collapse of one section of the Bay Bridge killed one person. Image: C. E. Meyer,
US Geological Survey, Digital Data Series DDS-29, version 1.2, image 16.

Figure 4.4
Journalists flocked to report on the intense damage to the Marina district of San Francisco. Image: A journalist works in San Francisco's Marina district after the October 1989 Loma Prieta earthquake, taken at Fillmore and Bay Streets by Nancy Wong, https://commons.wikimedia.org/wiki/File:A_journalist_works_in_San_Francisco %27s_Marina_District_after_the_October_1989_Loma_Prieta_Earthquake.jpg. This photo is licensed under the Creative Commons Attribution-Share Alike 3.0 Unported license. Some rights reserved. https://creativecommons.org/licenses/by-sa/3.0.

was no sense that members of the public might be the first responders to a disaster and thus be in an ideal position to provide immediate details.[46] Fourth, the USGS, the federal agency that housed and funded many seismologists, was not included in these disaster plans as a group that would communicate with the public.[47] Needless to say, the actual Loma Prieta response did not follow the ideal communication models laid out in the government disaster response plans. But this vision of the government as the authority did shape people's information practices, the information infrastructure, and how Californians made sense of the Loma Prieta earthquake.

After the Loma Prieta earthquake, the commercial media and government were in a complex dance, unsure of who was leading, as they produced and circulated information about the earthquake. While the government envisioned in disaster plans that the media would simply be a transmitter of state-generated information, many official disaster responders themselves relied on the media to learn about the disaster. Yet it was not as though the media ignored the government's words. On the contrary, the media sought out government officials to explain what happened. Responders at the regional and state levels of government, however, had difficulties getting an idea of what had happened on the day of the earthquake because their information networks were not operating as expected; these officials sometimes relied on the media to learn about what had happened. In turn, the media relied on the disaster response professionals for the official story, but also pressured the disaster responders for details that the disaster responders did not yet have. Both of these organizations reinforced each other's roles in shaping the public's knowledge of the earthquake.

Shortly after the earthquake struck, the California Office of Emergency Services set up an "Emergency News Center … staffed around the clock … by 50 public information officers from various State agencies."[48] As described by researcher Louise Comfort, the difficulty for state-level public information officers centered on waiting for the regional Operations Center report; the regional center, meanwhile, was waiting for counties to report, which were in turn waiting for cities to report, and the cities were too busy responding to the earthquake—putting out fires and helping people move into safe spaces—to make any reports.[49] The regional center could not do its job because there were no updates available regarding what was happening on the ground.[50] At sites such as the I-880 freeway collapse, initially the disaster responders at the site relied on runners to carry communications among disaster response professionals. Municipalities faced significant challenges in gathering information and providing succinct reports to the next regional level.[51]

In addition to these organizational communication challenges, the design of telecommunications infrastructures, which were supposed to facilitate communication between different government entities trying to work together to produce public information, created communication barriers. Emergency measures taken to decrease telephone line load blocked incoming calls from outside certain Bay Area phone area codes. This measure had unintended consequences: parts of Santa Cruz County could not get through to the Office of Emergency Services Region II Office in Pleasant Hill because people were calling from a blocked area code.[52] Retrospective assessment of the communication systems underscored that the telephone systems had issues because of the call volume, but that backup radio and cellular communication systems performed well.[53] As numerous people tried to get in touch with each other simultaneously, people thought the phone lines were broken even when they were working because people didn't know that they could wait for a dial tone.[54] Those who had access to cellular telephones had much better luck because the system had far fewer users relative to its capacity.[55] Overall, this telecommunications infrastructure was integral to enabling communication between some of the most affected communities and the regional center, and yet it was—by design—limiting possible event epistemologies.

Even when the telecommunications infrastructure was intact, the public information officers often did not have anything new to tell the press about the earthquake because the public information officers were getting updates from other professional disaster responders who were getting updates from the media. At some points in the immediate aftermath, the San Francisco public information officer got their stories from city workers who relayed information they received from television. The person in charge of public information in San Francisco struggled to obtain critical updates from emergency responders about what was happening in order "to feed ... the horde of press people."[56] The media's focus on the Bay Area along with the lack of attention to Santa Cruz County led officials in Watsonville, Santa Cruz, and Los Gatos to believe that the damage was so bad elsewhere, they were "on their own."[57] A 1990 Earthquake Engineering Research Institute summary of the socioeconomic aspects of the earthquake noted that "it appears that most organizations depended on the media to find out what was happening."[58] Predisaster planning documents described California's state-level emergency responders as responsible for pushing public information out as well as handling inquiries from the media and public. After Loma Prieta, however, these responders spent most of their time responding to journalists' questions rather than "using the media to communicate important information."[59] The Santa Cruz city manager reported that after Loma Prieta, "the media" completely had their own agenda.[60] Moreover, the disaster response plans had said that the public information officers were supposed to do "rumor control" and correct inaccuracies in media coverage, a task they were unable to do.

When reflecting on the wide variation in the initially confirmed number of deaths, the Santa Cruz postearthquake report suggests that the decision "whether or not to estimate casualties or damage" rested with the public information officer.[61] The source of the high casualty numbers at the I-880 freeway collapse was apparently both from disaster response officials and the California Office of Emergency Services, but the California Office of Emergency Services public information officers were likely repeating the number that they had heard from the public information officers at the site of the destroyed highway.[62] The design of the emergency plan—that all information would flow up from local offices to the state office—implied that the field office would give the same report as the Sacramento office, assuming that the Sacramento office believed the local report. This conflicted with journalistic efforts at fact-checking and sourcing information from two separate offices. Thus, even as journalists sought to confirm statistics from a variety of public information officers, the public information officers were all sharing a single story. Furthermore, inaccurate estimates such as the inflated number of people dead as 273 circulated; even four months after the earthquake, earthquake-related deaths were reported as between 62 and 66, rather than a single agreed-on number.[63]

News organizations and disaster response officials had an odd synergistic overlap when it came to telling the public what had happened. In postearthquake reports, disaster response officials complained that the media distorted the earthquake and the media complained that the disaster responders got important facts wrong. Yet the disaster responders set up venues for the public information officers to work with the media, and the media also frequently used government officials as sources.[64] The media were integral in bringing the voices of disaster response professionals to ordinary people, even if the media didn't always do what government officials expected of them.

Informing Earthquake Publics

Not only did the disaster response plans not account for their own limitations in creating authoritative information and the role of the media as an actor in telling stories about the earthquake, but the plans also had misunderstandings about the earthquake publics. These publics were not passively constructed by the public information infrastructure; instead, they participated in the public information infrastructure in order to meet their needs for details about what happened.

In San Francisco's 1988 *Multihazard Functional Plan*, the Emergency Broadcast System is denoted as the only radio communication system with frequencies specifically meant for government communication with the public. According to this plan, "The Emergency Broadcast System (EBS) will be used, to the maximum extent possible, for the dissemination of emergency information, advice, and action instructions to the general public."[65] The design of the Emergency Broadcast System relied on certain radio stations, designated as Common Program Control Stations, to receive the Emergency Broadcast System messages and send out a tone signal to other stations, which would then rebroadcast the messages. But the Common

Program Control Stations were not operational immediately after the earthquake due to a lack of electricity or other technical problems. Additionally, many radio stations chose not to monitor the Emergency Broadcasting System.[66] San Francisco mayor Art Agnos actually had his Emergency Broadcasting System messages translated into other languages, but there were conflicting reports about whether they were broadcast, though many reports agreed that the Emergency Broadcasting System did not do what people expected it would.[67] One group of researchers summarized its experience with the Emergency Broadcasting System after Loma Prieta: "In the case of an emergency, we expect that the Emergency Broadcasting System (EBS) will inform the public where to turn for necessary information. Unfortunately, the Emergency Broadcasting System failed to do so in the case of the Loma Prieta earthquake."[68]

Since the Emergency Broadcasting System, the main intermediary that the government envisioned being able to use to communicate directly with the public, was under utilized, citizens found other ways to obtain government-issued public emergency information. As discussed above, the disaster plans emphasized professional disaster responders communicating with the media, which would then tell the public what had happened. In Santa Cruz, however, according to reports by the city and local library staff, the library staff members assumed the role of local officials "providing information to the public."[69] The California Library Association is emphatic about the attention that citizens need after an earthquake in a 1990 recommendations document:

> The State-mandated (from the Office of Emergency Services) Emergency Management Plans (which every jurisdiction in the State is supposed to develop) *do not contain adequate provision for citizen information.* Instead, a Public Information Officer is designated. This person is supposed to handle inquiries from the press and other media and answer questions on behalf of the Incident Commander.[70]

The Santa Cruz Library, on request from the police department, set up and staffed a citizen information phone line starting on October 18, the day after the earthquake, and it functioned for at least two weeks afterward. The phone number was broadcast on the radio, and the library received calls from all over Santa Cruz County. Ideally, this service was meant to lighten the load on other city departments, such as the water department, which could only respond to questions in their domain and were already occupied with disaster recovery activities. Library staff responded to two different principal inquiries. The first were inquiries about individual well-being in which people from faraway places called to ask about the well-being of a particular person; "staff responded to [this] by locating the person (sometimes using a reverse phone directory to check with neighbors) and calling the person or the neighbor to send a message."[71] In the first three days after the earthquake, this type of call was most typical; staff fielded over two hundred inquiries. The second type of inquiry was from local residents seeking emergency-related information:

> Who to call for a building inspection, where to volunteer, where to send money, where the shelters were, when water would be restored, was the water safe, where to get meals, were the banks open, could a building on the Mall be entered, was there really a tidal wave coming, which schools were open, etc., etc., etc., etc. … The Library staff began compiling County-wide (as well as City) information, using all the resources at a librarian's command. The hardcopy data compilation was updated daily and staff soon discovered that other agencies were copying the material on their letterhead.[72]

Because the Santa Cruz Library staff members were responding to inquiries on an ad hoc basis, they weren't recognized in any official disaster response plan. Many people involved in the planned response did not know about the services that the library staff was providing; "the Library staff sometimes had trouble getting accurate information for use at the Service and several times got left out of the 'information loop' when conditions changed."[73] That the press would not supply citizens with the information they required was a scenario not foreseen in any disaster plans. The Santa Cruz city manager reflected that "our library took the initiative to provide these community information services, and they met a critically important, and unforeseen, need."[74]

Other city managers didn't see the need to create a separate function for non-media-oriented public information and put the responsibility of communicating directly with the public in the hands of the government's professional disaster responders, as originally envisioned. Oakland residents initially called 911 for information, until phone numbers for Oakland's Emergency Operations Center were broadcast that "could be called to obtain current disaster information," since "the demands for 'public information' cannot be underestimated."[75] The City and County of San Francisco's attitude was that the Emergency Command Center was a space reserved for professional responders, and staffers were caught off guard by calls from the public. Without the library or another city organization assuming the role of taking calls from the public, there was a lot of confusion in San Francisco as to whom the public could reach out to for information.[76]

The idea that earthquake survivors on the ground had some contribution to make to the ongoing narrative of what happened was almost never discussed in plans and rarely talked about in postearthquake reports. Nonetheless, there were many instances where people who actually experienced the earthquake had worthwhile experiences to contribute to responders. The media not only aired the stories that the disaster response professionals provided. In some cases, reporters treated this as an opportunity to hear directly from people and get a broad idea of the unfolding story. One reporter even claimed that he identified the area near Santa Cruz as the epicenter of the earthquake based on the lack of phone calls coming out of Santa Cruz County.[77] Obtaining a broad understanding of what happened was precisely the problem for professional responders. Yet people on the ground helped make the overall damage knowable for others far away.

Radio, the medium that many people turned to when the electricity went out, found phone calls from listeners especially useful. Even so, reporters often treated the stories from eyewitnesses as a last resort alternative to official narratives.[78] These grudgingly used eyewitness descriptions, along with the emergency plans, give the impression that the depictions by people

Figure 4.5
The *San Francisco Chronicle*'s caption read, "October 17, 1989—Watsonville, California: The night of the earthquake the Ceballos family of Watsonville sat by a BBQ fire to keep warm and watched a portable TV and stayed out of the house afraid of aftershocks during the Loma Prieta Earthquake." Image: Steve Castillo/*San Francisco Chronicle*/Polaris.

who had experienced the earthquake were somehow less valid. The *San Jose Mercury News* won the Pulitzer Prize for its unique reporting about the Loma Prieta earthquake; a 1992 study observes that the *Mercury News* sources were "much more likely to be scientists ... more likely to be eyewitnesses and less likely to be elected officials or representatives of government agencies."[79] Perhaps the scientists and eyewitnesses were the unique voices in the earthquake reporting otherwise dominated by public information officers.

Infrastructure Orphans

What I have described so far—the public information infrastructure that was constituted through the government in its production of public information, the mainstream media in their circulation of stories about what happened, and the English-speaking earthquake public—enabled knowledges about the earthquake that were mostly from the perspective of the English-speaking government and media. And while the government's plans might imagine that it would serve authoritative information to all those affected, it did not. The most powerful entities shaping the information order worked primarily in English. This government also assumed a middle-class, home-owning citizenry, leaving out many of the poorest Californians from its plans for recovery. Non-English speakers were sometimes elderly or with few resources, and were frequently the most in need of assistance after the earthquake. In Susan Leigh Star's terms, these were the orphans of the mainstream public information infrastructure.[80]

While government plans articulated the need for the dissemination of translations of public information immediately after the earthquake, this did not happen quickly, leaving some of the most vulnerable communities out of the loop, and reliant on the non-English-language media and civil society organizations for translated information.[81] In 1989, the Center for Immigrant Rights and Refugee Services estimated that two hundred thousand refugees lived in San Francisco, and many of these people were not proficient in English. Immediately after the earthquake, minority-language television and radio stations, usually smaller stations with smaller budgets, struggled to get back on air quickly. Until this happened, non-English speakers were left without "even the most basic earthquake information" including advice on basic safety, such as that people should not turn on a gas stove or light a match.[82] Some traumatized refugees fled the area altogether after the earthquake.[83] Local newspapers, television stations, and community groups provided translations of earthquake safety instructions. For example, a neighborhood newspaper, the *Tenderloin Times* published an issue several days after the earthquake that contained instructions in Lao, Vietnamese, Khmer, and Chinese for "How to Survive the Next One."[84]

Even when the immediate crisis lessened and people began to look toward long-term recovery, dominant public information infrastructures still excluded poor and marginalized communities. Recovery often meant applying to the federal government for some kind of aid to help find or rebuild shelter. The process of applying for aid—involving questionnaires or other paper forms—was another dimension of public information infrastructure that created orphans. The disaster response apparatus was meant to help those affected by disasters (usually people from the poorest and most marginalized communities), but as with the 1906 earthquake, disaster recovery programs were shaped by progressive ideology about who should be helped and how. As with the 1906 disaster, the job of the state recovery programs in 1989 was to restore people to their predisaster circumstances. Rather than divide aid equally between all those affected by the earthquake, members of the middle class were restored to their preearthquake living conditions, whereas the poor or transient were returned to their previous difficult living conditions, with little help from the government.[85]

The way that FEMA defined an "aid recipient" frequently did not fit the description of the people who needed actual housing aid.[86] For instance, many poor people left homeless by the earthquake didn't have the paperwork that FEMA needed and were not helped initially.[87] The preearthquake homeless population was also almost entirely excluded from the disaster response process.[88] Many postdisaster reports echoed the idea that the most marginal and vulnerable members of society suffered the most and most relied on aid.[89] Marginalized people—such as non-English-speaking refugees, the elderly living on fixed incomes, and people in short-term or low-quality housing—essentially didn't fit the category of people needing housing that FEMA was trying to address. And the public information infrastructure constituted by aid forms reified the exclusion of the extremely poor.

Spanish Speakers in Watsonville
In the aftermath of the disaster, those orphaned by a theoretically inclusive public information infrastructure often turned to alternative ones, but these infrastructures were not simply replicas of their English-language counterparts. Spanish speakers in 1989, like Chinese speakers in 1906, produced alternative information infrastructures. These provided the Spanish-speaking population access to resources in a manner different than for the English-speaking population. The alternative public information infrastructures were constituted by the ways of accessing disaster aid, community-based

health organizations, translations of government-produced public information, and Spanish-language media outlets. The rest of this chapter explores
how the Spanish-speaking population in the Bay Area, especially in the
hard-hit area of Watsonville, constituted an alternative earthquake public,
and the various overlaps among different public information infrastructures.

The city of Watsonville, located near the earthquake epicenter in Santa
Cruz County and home to over eighteen thousand Hispanic residents (61
percent of its population), lost 8 percent of its housing (642 of its 8,100
housing units).[90] This represented a significant housing loss; there was less
than 1 percent of housing available before the earthquake, and a number
of people became homeless.[91] Watsonville was also historically a location
for Latino farmworker organizing and advocacy. A few months before the
Loma Prieta earthquake, Latinos had won a court case that would enable
more equal representation in city elections; despite being a majority of the
population, there was only one Latino on the city council, and he was
thought to represent the interests of developers, not the community.[92]
Scholars argue that the Latino community's history of resistance helped
contribute to making Watsonville a site of "contentious collective action."[93]
Those left homeless fought at various times with the city, the Red Cross,
and FEMA over the accessibility of government services.[94] Members of the
Spanish-speaking community navigated the public information infrastructure set up by disaster response organizations, including aid forms not
designed with the multilingual population of Watsonville in mind. These
forms were integral to accessing the aid made available through the federal
government via FEMA, which finally, a week after the earthquake, formally
announced that all residents could apply regardless of immigration status.[95]
Though the aid forms seamlessly constituted part of the public information
infrastructure for the English-speaking earthquake public, they represented
a barrier for those outside it.

Spanish-Language Media

In both Watsonville and across the greater region, the Spanish-speaking
population, and particularly the Spanish-only-speaking population, was
left out of mainstream Loma Prieta disaster communication. Indeed, Spanish speakers found it challenging to make sense of the earthquake using
the broadcast media. One Bay Area newspaper, the *Bay Guardian*, ran the
following headline: "Earthquake Night News: For English Speakers Only."[96]
In a 1992 report by Federico Subervi-Vélez and colleagues, *Communicating with California's Spanish-Speaking Populations: Assessing the Role of the
Spanish-Language Broadcast Media and Selected Agencies in Providing Emergency*

Services, there is special attention paid to Spanish-language radio and television stations. The report describes the television and radio components of the alternative public information infrastructure that Spanish-only speakers might have had access to as they made sense of the earthquake.[97] Radio was important to this earthquake public; on the day of the earthquake, 62.5 percent of Hispanics found the radio to be the best source of information, and 36.1 percent found the television to be the best source of information. "Hispanic Californians" were more likely than non-Hispanic Californians to "turn on or find a/another television or radio to get more information," suggesting that it may have been difficult for Spanish-only speakers to find accessible information sources.[98]

While Spanish-speaking communities in large cities such as San Francisco or San Jose had multiple Spanish-language media options, smaller towns like Watsonville usually had only one local radio station that broadcast Spanish-language content for only a few hours per day. Furthermore, Spanish-language stations in San Francisco and San Jose had more resources: they had reporters out in the field with cellular phones, and were able to field calls from distressed Spanish speakers to answer questions about "specific roads, schools, water conditions, and places to go for governmental assistance."[99] On the day of the earthquake, the part-time Spanish-language Watsonville station, KOMY-AM, had "English-language news and information ... transmitted along with brief translations into Spanish."[100]

Spanish-language radio stations acted to help translate government public information for Spanish speaking communities, because most government information was offered only in English. Even the most public-facing government information system, the Emergency Broadcast System, was only run in English. Though the Emergency Broadcasting System had symbolic importance as a structure dedicated to government agencies communicating with the public (whereas the public information officer or Joint Information Center were dedicated to communicating with the media), it was not, as discussed previously, used to its fullest potential during the Loma Prieta earthquake. Moreover, the design of the infrastructure for the deployment of Emergency Broadcasting System messages did not accommodate non-English messages. Common Program Control Stations were all English-language stations, and the messages sent out would all be in English. The Emergency Broadcasting System represented the government's most concerted effort to put out emergency public information directly to the people, but it did not consider non-English speakers.

Reports before Loma Prieta on other California earthquakes in the 1980s had already identified that the government had difficulty communicating with the Spanish-speaking population.[101] The vision of the "response" to the earthquake in disaster response planning documents theoretically included non-English speakers. For example, the state of California's 1985 *Multihazard Functional Planning Guidance* instructs that the state should "provide EPI [Emergency Public Information] in foreign languages as required." A blank form that was meant to be filled in with contact information for "Emergency Public Information Staff" supplied the following designation: "* (S) following name denotes Spanish speaking."[102] Furthermore, blank tables were provided for "translator services," to be filled in with names along with contact phone numbers and addresses. Despite these gestures in the planning documents toward serving a multilingual audience, however, the provided "sample radio message" was given only in English. Though there are indications in disaster preparedness materials from that period that translated materials would be made available in other languages, my examination of disaster preparedness plans and materials for provisions concerning non-English speakers makes it clear that follow-through for access to non-English-speaking residents did not exist.[103]

Across the Bay Area, Spanish-language stations translated the English-language media along with Red Cross and government announcements. Additionally, in the weeks after the earthquake, FEMA used Spanish-language radio stations to tell Spanish speakers about their right to apply for aid.[104] Although the Spanish-language stations did spend time translating material from English-language sources and monitoring English-language stations, they did not run exactly the same translated content as the English-language stations. Barbara Garcia, director of Salud Para La Gente, a community-based health organization, used the Spanish-language media to draw attention to Watsonville, because she felt that the city of Watsonville was downplaying the damage and ignoring the welfare of Latinos.[105] This media spotlight was not always welcome in some cases. Watsonville city employees felt that the media unfairly promoted the agenda of those who had fled their homes and settled in public parks without official sanction from the formal relief apparatus: "Media attention fueled the demands and encouraged entrenchment rather than relocation to Red Cross shelters."[106] Since the agenda of the Watsonville local government did not meet the needs of those in Watsonville requiring aid, many of whom were Latino, the Spanish-language media sought to provide attention and resources to the situation.

Community-based leader Garcia and Spanish-language stations focused on the effects of the earthquake on the local Latino communities: "Calls from the audience provided invaluable information about particular experiences and problems. For example, people told us about local gas leaks, broken water pipes and structures that had collapsed or been damaged."[107] A few Spanish-language radio stations and both of the Bay Area Spanish-language television stations ran call-in shows during which people could ask questions of Spanish-speaking emergency professionals, geologists, or psychologists.[108] A few Spanish-language stations also facilitated sending messages about people's well-being to friends and family far away. For instance, KIQI-AM in San Francisco, working with its affiliates in Latin American countries, was able to record messages from those who were in areas affected by the earthquake and play them in South American countries.[109] This creative use of Spanish-language public information infrastructure helped serve the needs of a specific community.

Alternative Public Information Infrastructure and Accessing Aid
Public information infrastructures were not only constituted through different media to help people make sense of what had happened immediately after the earthquake but were also constituted through the forms and other types of documentation involved in getting aid after the earthquake. In the absence of accessible public information after the earthquake, Spanish-speaking residents in Watsonville turned to an existing network of Latino organizations. These organizations were an especially important resource for those who were displaced from their homes and living in the public parks.[110] Whereas the organizations that supported the Latino community in Watsonville were utilized before the earthquake, unfamiliar government disaster aid programs were not. Salud Para La Gente worked to translate English-language media as well as set up a clinic and "information booth" at the central Watsonville plaza, offering translation services for help understanding government documents such as the colored building tags.[111] After the earthquake, building inspectors tagged buildings with green (safe to enter and occupy), red (not safe to enter), and yellow (entry limited to authorized personnel) in order to indicate structural safety assessments. The explanations for the tag colors were posted in English only, though, and "even the meaning of green-tagged buildings [could] be disputed given a context in which residents do not necessarily trust the opinion of those who have the authority to assess the habitability of buildings."[112] In this sense, Salud Para La Gente, already known and trusted in the community, was part of the ongoing process of producing an alternative public information infrastructure in which Spanish speakers produced knowledge about the earthquake. The Spanish-language media were instrumental in helping spread public information, and community groups such as Salud Para La Gente were part of the process of public information infrastructure for Spanish speakers.

In an oral history interview, Garcia remembers that the group set up an emergency treatment center near its office on the Watsonville city plaza within minutes of the earthquake to assist the injured. It was open twenty-four hours a day after that to assist people living in Callahan Park, one of the unsanctioned parks, with medical issues and translation needs.[113] Three days after the earthquake, Latino leaders held a rally on the Watsonville city plaza, where "community groups set up information tables at the rally, while the city did not provide any."[114] Garcia points out that FEMA and the Red Cross planned to work directly with the city and county. Yet the city government of Watsonville was not representative of the people who lived there; despite having over 60 percent Latinos in the Watsonville population, the city had one Latino on the city council, as mentioned earlier.[115] Community groups were crucial for reaching out to the Latino community, because the city and county had little involvement with Latino communities prior to the earthquake.[116] Watsonville's postdisaster report fully acknowledges that leaving out representatives of the Latino community precipitated the struggles between earthquake survivors and disaster response officials.[117]

FEMA opened Disaster Assistance Centers in Santa Cruz County where people could go and apply for disaster relief on October 22, 1989—five days after the earthquake.[118] Unfortunately, these centers were not easily accessible to residents of Watsonville, particularly for those without a car.[119] When people moved (or were forced) out of temporary shelters, they were eligible to apply to FEMA for rental assistance, but the forms were often only available in English.[120] These forms constituted part of the public information infrastructure implemented by government and quasi-government organizations—but initially, this infrastructure was exclusive to English speakers.[121] Spanish- and English-language media both brought attention to the linguistic issues for aid seekers. As a result the Red Cross adopted more culturally appropriate services, eventually employing a bilingual caseworker, and the city of Watsonville appointed members of the Latino community to earthquake relief groups.[122] According to the 1990 State/Federal Hazard Mitigation Team postearthquake report, the California Office of Emergency Services and FEMA did eventually provide equal access to all.[123]

Language was not the only factor causing difficulty in the ability of Latinos to access the system of FEMA aid. The multigenerational families or multiple families living under one roof did not fit into the model of a "household" that FEMA envisioned in designing the forms, and many Latinos found it difficult or impossible to obtain the proof of residency records required to apply for certain kinds of aid.[124] Thus, some earthquake publics were excluded from the formal disaster response apparatus that was set up by the state not only for linguistic reasons but also because of culture and class. The earthquake publics constituted through Salud Para La Gente and Spanish-language media, and the public information infrastructure that included the state and English-language media, overlapped as well as having moments of separation and exclusion.

Conclusion

This chapter looked at the public information infrastructures that Californians participated in as they worked to make sense of the Loma Prieta earthquake. I focused on the government's claim to informational authority both embedded in disaster response plans and practiced in simulations. The government built notions of an earthquake public into the plans involving public information. This earthquake public would be communicated with via the media, which would broadcast the public information provided to them by specially appointed disaster response officials. The public was assumed to be English speaking and living with a single family in a residence. The limits of this imagined earthquake public, however, were revealed during the earthquake response.

Although there was a wider variety of broadcast media available after the 1989 earthquake, there was also interdependence between newspaper and television organizations. The use of backup electricity for so much of the information infrastructure in 1989 represents a key shift from the information infrastructure in 1906 or 1868. The reach of public information infrastructures, particularly national television stations, meant that the audience for the earthquake was the entire United States or even the world. The commercial media was interested in providing a vision of the earthquake for a national audience, focusing on San Francisco and the most spectacular destruction of life on the I-880 freeway collapse in Oakland, at the expense of hard-hit areas such as Santa Cruz and Watsonville. When possible, earthquake survivors immediately turned to radio and television to learn what happened, but as in the 1868 and 1906 earthquakes, the nature of the reach of infrastructure meant that many of the stories that circulated about the earthquake were not for the people actually affected by it.

The Loma Prieta earthquake shows that the government's vision of the dissemination of public information via the corporate media was flawed. The government's notion of how it would collect information didn't work because at the local level, people were concentrating on the work in front of them and didn't have the resources to spend time reporting to higher levels of government. The disconnect between the regional and state levels of government and the municipalities trying to respond to what was happening on the ground was further reinforced by the technical infrastructure that didn't allow the receipt of incoming calls from Santa Cruz County. The government itself learned about the earthquake from the media. The government's response was therefore shaped by the media's emphasis on San Francisco. Meanwhile, some communities came up with creative work-arounds to get past the shortcomings of the disaster response plans. Santa Cruz librarians staffed a community information phone line for the public. In Watsonville, Spanish-language media and community-based organizations produced an alternate public information infrastructure to support Spanish-speaking earthquake publics. The different public information infrastructures supported overlapping groups of people in generating knowledge about the earthquake. That is not to say that these different earthquake publics were equivalent in their power within the information order; simply the presumption that the government was supposed to serve everyone, but didn't, establishes a specific ordering among the public information infrastructures.

Despite issues with both the media and government in their conception of earthquake publics, the media frequently used the government as a source for the stories. The disaster response apparatus and media had a symbiotic relationship in which they both reinforced each other's centrality. The material public information infrastructure in 1989 appears more varied than in 1906 because of the added technologies such as television and radio, but in fact in the days after the earthquake, the mainstream media and government seemed to reinforce a singular narrative. However, there is more work by academics, librarians, and journalists documenting the experiences of earthquake publics that made alternative public information infrastructures, making them more visible to people today.

The job of planning effective public information infrastructures is always a difficult task undertaken by various entities in the government that attempt to shape earthquake knowledge. In 1989, the diversity of the US populace was difficult for the government to appreciate in its work on disaster response plans. Furthermore, the information order was dominated by a number of media companies interested in describing the earthquake in terms that were appealing to their audiences—their customers—across the country rather than the government. The bureaucratic apparatus for disaster response and the media were both important in constituting post–Loma Prieta earthquake publics. These relations remain relevant to contemporary disaster response planning and the information order of today.

5   Today: Government Disaster Response Meets New Media Platforms

My first thought was to update my status on Facebook. ... My phone beeped and I checked it, thinking it would be from my wife. Instead, it was a half-dozen posts from friends on Facebook in Japan relaying either their own "shaken-but-okay" statuses, or worried posts from [friends in] the United States wanting details ...

More friends checked in while the earthquake was upgraded from 8.8 to 8.9 and the tsunami was reported as ten meters high in places. Just numbers on a phone screen.

—Joel David Neff, Takanezawa, Tochigi

Above all, my greatest anxiety is caused by the radiation leak from the Fukushima Daiichi nuclear plant. I think that the biggest problem has been the transfer of information. The only thing I can praise the Tokyo Electric Power Company for is the rapidity and accuracy of reporting radiation levels to the public. ...

How is the condition of the nuclear plant going to affect us, how far is the risk going to spread, and what is the possibility of this happening?

—Miho Nishiro, Abiko, Chiba

I opened this book with a postearthquake informational experience in Japan that might feel familiar to a contemporary reader.[1] Though I then moved back in time and across an ocean to examine information orders from disasters in Northern California, I find that when focusing on the present day, these remembrances of the 2011 Tōhoku earthquake and tsunami highlight the contemporary technologies and institutions involved in mediating the disaster. As with all the cases I've explored, there are elements of the postdisaster information ecosystem that span time and geography, while others are specific to a context. To many readers, the technologies that are woven through Joel's story told above of the Tōhoku earthquake and tsunami—Facebook, for example—are familiar. Whereas without knowledge of Japanese institutions such as the Tokyo Electric Power Company, the organizations in Miho's story and their trustworthiness as informational authorities might not be familiar. Drawing on my analysis of the historical earthquakes, this chapter looks at the potential public information infrastructure if an earthquake were to happen today in the Bay Area in Northern California. What are the institutions, forces, policies, and technologies that shape today's postdisaster information landscape?

Each of the previous chapters makes it clear that the specificity of when and where an earthquake happened was crucial to what followed. The 1989 earthquake importantly happened during a World Series that featured Bay Area baseball teams; many people who might have been on the roads during rush hour—roads that were destroyed during the earthquake—were instead watching television. The earthquake epicenter was near Loma Prieta, not, as the 1868 earthquake was, centered in Hayward in the populous East Bay. Furthermore, the specific populations and infrastructures that were shaken shaped the entire arc of what happened next. So it is difficult and perhaps unadvised to guess what might happen after a Bay Area earthquake today. But I can consider the information order that is currently in place today and what we know about the particular event epistemologies it produces in other disasters.

A notable trend throughout this book is what I call the *bureaucratization of disaster response.* In the earthquakes discussed early on in this book, the government did not plan for disaster response. In earlier disasters, especially in 1868, earthquake publics struggled with the state's role in disaster response. In 1868, no plan existed for how the government would react to an earthquake. The follow-up to the 1906 earthquake and fire saw a larger local government response to the disaster, but again, it was not based on a planned and publicly vetted disaster response process. The 1989 earthquake illustrated the degree to which this bureaucratization of the government disaster response had been put in place throughout the Cold War. The 1989 earthquake included planned informational responses—particularly in the production of public information.

This bureaucratization of disaster response is also important to contemporary event epistemology. As I explore below, today's disaster response plans imagine a postdisaster space and often contain explicit instructions for how disaster response professionals should communicate with "the public." Conceptions of (a singular) "the public" in contemporary disaster response plans aim to be inclusive in their outreach. The instructions for creating public information still aim for the government to be the informational authority. Yet, the disaster plans also imagine how disaster response professionals should use information generated by citizens. Additionally, contemporary plans contain fairly explicit expectations for what US residents believe to be the government's responsibilities after an earthquake—responsibilities to make people safe, rebuild, manage a disastrous situation, and circulate documents that help with these tasks.

Past chapters of this book show that event epistemologies are formed and limited by whatever means are familiar and available. From this perspective, in addition to concentrating on the government's emergency response plans, it is useful to consider established public information infrastructures. As the chapter about the 1906 earthquake demonstrates, companies with resources and established work practices frequently have the means to attempt to find work-arounds, even in the face of physical breakage. Today, well-resourced companies may be able to use their deep pockets to provide continuous services and minimize breakdowns. Many people—especially the elderly and impoverished—lack resources such as regular Internet access and may not benefit from this. While an increasing number of people have cell phones, the cost of accessing the Internet on these devices remains inaccessible to many. Technologies used by earthquake publics in 1989—radio, paper newspapers, and television—are still key sites today where earthquake publics are constituted and disasters are made sense of. But as media theorists have noted, the "old media," so critical in the earthquakes described in this book, has morphed and converged with the "new media"; the distinctions between radio and the Internet can be difficult to parse if you listen to your favorite radio stations online. People find newspaper stories via Twitter and get updates about loved ones through Facebook. Examining public information infrastructures with these dominant media company platforms in mind allows for the inclusion of many different kinds of information practices. This focus remains limited, however; it is not encompassing of all information practices and all people.

Instead of analyzing what happened in a specific historical moment as I have in other chapters, I look at the disaster planning materials and artifacts that we have in place today. This includes not only government planning documents but also artifacts from social media corporations specifically intended to support users after a disaster. Similar to government disaster plans, imaginings and past experiences of crisis are inscribed in technologies. These platforms and disaster plans might actually shape what is to come, but also reflect the present in which they were made. Despite the future orientation, I rely on the same techniques as I have in previous chapters: publicly available documents that can get me as close to imagined practices as possible. I analyze the present through a close reading of current disaster response plans, and analysis of information and communication technologies intended to be used by citizens after a disaster. In this chapter I read present-day San Francisco earthquake disaster plans at the city, regional, state, and federal levels to understand how information and communication practices are envisioned for earthquake publics. I look at disaster response plans in contrast to the ways in which imagined communication practices are inscribed in the design of information and communication technologies.[2] The information practices inscribed in the disaster response plans and ideas that have underpinned social media sites can be understood as oppositional: professional responders seek to create and circulate an authoritative document, and social media authority rests on the multiplicity of documents created by the "many."

The Information Production Dialectic

On the one hand, the public information infrastructure of today is conceived of in terms of the production of documents by the many—the masses of Google, Twitter, and Facebook users, whose voices are broadcast far beyond the streets from where they access these platforms. The traditional argument is that before the Internet, there were fewer venues for broadcasting a message widely—for example, libraries, radio, television, and newspapers. These institutions produced and culled content, and decided whose voices got heard. The mainstream media's power to decide who gets broadcast has been shifted; everyone with access to the Internet may technically circulate documents in the public sphere, even though many of these documents may never get read. And through "the power of the many," comes an imagined authoritative voice illuminating what is really happening after a disaster. In theory, these new media platforms allow anyone to tell their story of what happens after a disaster, shifting how information authority is practiced.

On the other hand, the government response plans describe hierarchical organizational systems, such as the Incident Command System (discussed in more detail later in the chapter) for producing authoritative information to be distributed to earthquake publics. Though disaster response plans and sociotechnical media platforms are both used to produce information about a disaster, one could characterize bureaucratic technology and information technology as forming a dialectical relationship.[3] Taking these models of information production to their extremes, the government disaster plans articulate a vision of a unitary centralized voice, and information technologies such as Twitter represent a potential cacophony of voices. The idea that the government can produce a kind of unitary knowledge is in dialectical opposition to the plurality that the new media technologies supposedly enable.

Yet when I examine the role of information practices and technologies in disaster planning, seemingly oppositional forces are intertwined in symbiotic ways. People seek to reach government disaster response organizations using social media, government disaster response organizations use social media to reach the earthquake publics they are trying to help, social media companies make products to account for people after a disaster, and researchers build tools to help government disaster response organizations attempt to use social media information in their response activities. Information technologies can, and are, being integrated into the centralizing information practices described in disaster response plans. Social media platforms fit in well with the government's disaster plans because they can be integrated into the hierarchical and centralized Joint Information System, explored later in this chapter. Twitter data could be seen as an easy source of information to be aggregated for disaster responders making decisions. And platforms like Twitter can also give the government a venue for directly distributing its "public information." If Twitter is used to create the government's overview of the disaster, however, which voices will be highlighted and which will be downplayed?

The government maintains a presence on crowdsourcing websites such as Twitter and Facebook, and uses these platforms to broadcast its messages. Government accounts can get a disproportionate amount of the traffic on Twitter in times of crisis.[4] In some sense, the government, which uses Twitter, now has more control than ever over its communications with citizens. Disaster plans used to rely on mainstream media outlets to circulate the governments' messages, and disaster planners would simply hope that people would turn on their radios or televisions to receive information via the Emergency Broadcast System.

This chapter explores how this dialectic plays out in the production of authoritative information about a disaster. First, I examine how the government produces public information and creates situational awareness through a number of organizational innovations, namely the Incident Command and Joint Information Systems. I also look at how the government utilizes various social media and other mobile technologies in order to make its public information available to citizens. I then turn to disaster response activities that are envisioned by social media companies. Using these examinations of various platforms and plans, I elucidate the contemporary information orders that might shape the experience of an earthquake for San Franciscans.

Government's Information Authority

Contemporary US disaster plans imagine the techniques that the government, as part of the information order, should use to produce information. These disaster response plans themselves are a form of information infrastructure in the sense that they portray and are representations of bureaucratic technology. In other words, the disaster response plans describe the actions that professional disaster responders should take to produce public information. And the plans themselves are a material representation of disaster information practices. While these bureaucratic technologies in no way constitute the whole of the information order after a disaster, they are worthy of examination because experience tells us that disaster plans critically shape the government's actions. Understanding how government-run elements of public information infrastructures function is important, but requires discussion of the complicated organizational concepts described in the disaster response plans.[5] One of the points that my analysis will make clear throughout this section is that these government organizations, which are aspects of the information order, are complicated entities, which make use of specialized terminology and organizational structures.

In the context of the United States, disaster plans give an idealized picture of government involvement in postdisaster information infrastructure. Plans explain how to both preserve the past and make the future. The plans preserve the past by, in some sense, assuming that the goal is to help people return to a predisaster existence, and the plans attempt to make the future by guessing what needs to be done after a disaster. These plans imagine a postdisaster space of communication and often contain explicit instructions for how disaster response professionals are to communicate with earthquake publics.[6] I look at two key activities that the government's bureaucratic technology takes on: producing public information to inform and instruct the public, and creating a situational awareness that guides the state's activities.

The organizational schemes represented in the disaster plans describe bureaucratic technologies that are brilliant in their uniformity of vision across plans. Today, these plans use the Incident Command System, based on the idea that all agencies responding to a disaster will use the same terminology and organizational structure, providing interoperability to organizations that may not have previously worked together. The Incident Command System is a program that intentionally implements "institutional isomorphism."[7] That is, the idea is that everyone who shows up to respond to a disaster in a professional capacity understands the terminology and organizational structure that will be used by all disaster responders.[8] The Incident Command System includes the concept of a "Unified Command," such that there is only one path up a chain of command, every person has only one supervisor, and there is a single "incident commander" in charge of all the decisions about operations at the incident site.[9] The Incident Command System is designed to be applied to all disasters, regardless of size, and can expand or contract to include various "emergency support functions," or instructions for extra organizational activities in a specific area, such as "transportation" or "search and rescue," which may or may not be needed in a particular response.

The Unified Command organizational structure attempts to centralize information within the government. In documents detailing the Unified Command system, information is a *thing* that just needs to be channeled properly—in this case, from the site of disaster to local government officials, then up to the state and national governments, and finally out to the public.[10] Although there are several places where the actual making of information is articulated, in general information is supposed to "flow" upward through the hierarchy to the person in charge—the incident commander. This *all-hazards approach* to disaster planning means that the basic national disaster plan to respond to a tornado in Missouri is the same as a plan to respond to an earthquake in San Francisco.

The Incident Command System is part of a set of organizational practices guiding US disaster planning, encoded in the US Department of Homeland Security's *National Incident Management System* (NIMS) documents.[11] In California, the *Standardized Emergency Management System* (SEMS) guides local disaster response. Although NIMS was adopted nationally after the September 11, 2001, terrorist attacks, it had been in development since the 1970s in the context of wildfire fighting in California. SEMS predates NIMS, although today, all local disaster response operations must conform to NIMS.[12]

The federal response to a disaster is supposed to dovetail that of the states, and the states support regional and local governments' work. In the case of San Francisco, which is both a city and county, the state government for California would only step in if San Francisco did not have the capacity to respond to a disaster and San Francisco's mayor requested help. And the federal government would only step in if California's resources were overwhelmed and the California governor requested help. Nevertheless, all levels of response are supposed to conform to the Incident Command System outlined in NIMS, regardless of whether the federal government is involved or not. Local plans must conform to the federal NIMS structure.

The system for processing messages outlined in these plans at different levels of the organization is in the interest of creating authoritative information. The logic of the disaster plans requires that all the information be centralized such that there is a single, consistent message to the (singular) public.[13] The primary system for producing information and circulating it to the appropriate decision maker, specified by the Incident Command System, is the Joint Information System. It is composed of multiple Joint Information Centers located within Emergency Operations Centers at different levels of government (e.g., federal, state, regional, and local).[14] The Joint Information Centers are supposed to advise the Incident Command System and other agencies; act as a "central point of contact for all news media"; develop, recommend, and execute "public information plans and strategies"; and "control … rumors and inaccurate information that could undermine public confidence in the emergency response effort."[15] Public information officers staff the Joint Information Centers and are part of the Incident Command System. The public information officers are "responsible for interfacing with the public and media" as well as other government agencies.[16] Within the Joint Information Centers, there are formal processes for "releasing information" to the public.[17]

One of San Francisco's disaster response plan documents, *Emergency Support Function [ESF] #15: Joint Information System Annex*, provides extensive details about how communication with citizens is supposed to work.[18] In San Francisco, the response plans outline that the Emergency Operations Centers will house the Joint Information Centers, and within each Joint Information Center is the Joint Information Section (not to be confused with the overarching Joint Information System).[19] All the portions of this organization—which together make up the Joint Information System— are supposed to process information to inform the government's disaster response as well as make "public information" in the context of the Incident Command System.

Contemporary federal, state, and local disaster response plans describe a hierarchical structure of information processing within the Incident Command System that aims to produce a definitive narrative for the earthquake public about what has happened and what to do next. All these disaster plans indicate that the government takes its role in being an informant of the earthquake public seriously. These plans describe a multilayered public information infrastructure. From the Joint Information System, Joint Information Centers, and in San Francisco, Joint Information Sections, to the policy of clearing information through the Incident Command System before releasing it, all aspects of these plans are intended to ensure that the government emergency professionals are capable of producing authoritative information about a disaster The next two sections elaborate on two specific information practices embodied in these plans: situational awareness and public information.

Producing Situational Awareness

One of the key informational activities that the government engages in is producing *situational awareness*. In the case of disaster response plans, situational awareness involves centralizing records associated with incidents as well as informing disaster response both broadly and specifically in the area of public information. Situational awareness is thought to be a state of understanding the implications and context of a disaster such that one can make decisions about what to do next. The 2008 *California Catastrophic Incident Base Plan Conops* (concept of operations) describes both the importance and difficulty of obtaining situational awareness about a disaster:

> In any major incident, the degree to which key decision makers at all levels of government and within interagency structures are able to gain and maintain a situational awareness on the scene determines, to a great degree, their ability to anticipate requirements and provide appropriate resources. Real-time situational awareness also facilitates timely and knowledgeable information-sharing with elected and appointed officials, the public, and the media.[20]

Situational awareness, which has its roots in military aviation psychology, informs the government's actions and efforts at producing public information.[21] The Joint Information System and other bureaucratic organizational arrangements created to produce information about disasters have been dedicated, in part, to addressing the problem of creating situational awareness.

In the disaster plans, situational awareness is achieved through collecting, analyzing, and summarizing reports related to the disaster. Efforts to "develop situational awareness" at a regional level, for instance, frequently take the form of aggregating findings from the affected areas. Sources for situational awareness range from disaster response coordination materials, reports from state and federal agencies, and maps and models, to on-the-scene news gleaned from social media platforms and traditional news sources.[22]

The US Department of Homeland Security's *Emergency Support Function #15: Standard Operating Procedures* describes how situational awareness is obtained not just from government reports but also from monitoring social media: "During an incident, ESF #15 should use publicly available social media sites for situational awareness, and should search on appropriate keywords, hash-tags, and other search terms on digital channels to find information for situational awareness."[23] During Hurricane Sandy in 2012, "more than 15 [FEMA] staff members" supported the agency's social media operation, which revealed information about Hurricane Sandy's impact that contributed to FEMA's situational awareness.[24] Later in this chapter I explore how social media companies are entwined with the state's vision of situational awareness.

Situational awareness is a goal that is invoked often in new plans for disaster response, particularly around the practices of information and communication. It is a technique that serves people in power, who are often at a distance from a disaster, in making decisions about how to respond. It also legitimizes choices around the mobilization and distribution of resources. After Hurricane Katrina in 2005, federal postdisaster analysis blamed a lack of situational awareness for what was widely agreed to be an appalling disaster response: "The lack of communications and situational awareness had a debilitating effect on the Federal response."[25] In afterdisaster reports, situational awareness along with the information supposedly underpinning it is a way to call attention to what people understood to be happening at the time of the disaster, and serves as a target of blame for poor decisions made due to limited or incorrect information.

Public Information   Within the *State of California Emergency Plan*, the idea of *public information* as an information practice mostly casts people who experienced the disaster in the role of receivers: "Public information consists of the processes, procedures and systems to communicate timely and accurate information by accessible means and in accessible formats on the incident's cause, size and current situation to the public, responders and additional stakeholders (both directly affected and indirectly affected)."[26] The plan elaborates that the government is the producer of public information and "responsible jurisdictions disseminate information about the emergency to keep the public informed about what has happened, the actions of emergency response agencies and to summarize the expected outcomes of the emergency actions."[27] While the state increasingly looks to different disaster publics for situational awareness, the plans still treat the government as primary informers of citizens.

The newest disaster response plans attempt to produce more inclusive "public information" such that a wider swath of "the public" can understand it. The *National Planning Frameworks*, five disaster preparedness documents published by FEMA, all emphasize the challenges of communicating with many different kinds of people.[28] For example, the *National Mitigation Framework* instructs local agencies to

> use social media, Web sites (e.g., Ready.gov), and smartphone applications, as well as more traditional mechanisms, such as community meetings or ethnic media outlets, to inform the public of actions to take to connect preparedness to resilience. Information and messaging should ensure effective communication with individuals who have disabilities or access and functional needs, including those who are deaf, hard of hearing, blind, or have low vision, through the use of appropriate auxiliary aids and services, such as sign language and other interpreters and the captioning of audio and video materials.[29]

For the government, the focus of public information production is getting the information into an appropriate format for different audiences. The next sections examine different programs attempting to produce public information and circulate it broadly.

Suggested Information Releases At the federal level, one illustration of the efforts at including multiple earthquake publics is found in an appendix to FEMA's *Emergency Support Function #15: Standard Operating Procedure* plan. The document describes how all "relevant incident content" will be "aggregated and curated" on USA.gov and gobiernoUSA.gov, a Spanish-language site, "with the same mandate to provide information to people with Limited English Proficiency."[30] Furthermore, the content will then be reposted on many sites and platforms.

The city and county of San Francisco envision reaching out to civilians after a disaster in an impressive variety of ways. Throughout the city and county's plan, there are details about how an accessible communications specialist will aid in disseminating information such that it is comprehensible "by people with disabilities and others with access and functional needs": AlertSF, a city-to-subscriber text message system; 311 ("operators taking calls from the public"); television, radio, print, and social media; "partner/stakeholder lists"; "printed information distributed door-to-door"; an Outdoor Warning System; the Emergency Alert System; and "on-scene loudspeaker announcements."[31] An assessment by the Bay Area Emergency Public Information Network, however, notes that San Francisco, an extremely wealthy city, is among the most prepared of counties in the region with resources available for contacting the residents of the city, yet they say that it is unclear if the San Francisco government will be able to reach out to multiple linguistic communities—even though the technology exists to do so.[32]

Even when not focused explicitly on inclusive information dissemination, San Francisco's disaster planning materials contain scenarios that explain how the state envisions public information being produced and circulated.[33] These scenarios are used to imagine activities that might take place after an earthquake, and assert the city's responsibility to use multiple communication channels to provide people with information about how to be safe and recover.[34] While public information provisions are not a priority in earthquake planning—lifesaving activities are—the plans imagine that within the first four hours after an earthquake, the disaster response operation will "begin public information messaging regarding recommended personal protective action, safe congregation point, and community assistance needed."[35] Templates illustrate what kinds of public information the government imagines itself producing. In the interest of standardizing messages, the *San Francisco Emergency Response Plan: Earthquake Annex* includes some "Suggested Post-Earthquake Public Disaster Information Releases" for dissemination during the first seventy-two hours, including statements instructing people on personal information practices.[36] These documents try to anticipate the earthquake publics and their needs. The disaster response plans attempt to use many approaches to try to reach a variety of San Francisco denizens while also being crucial to the formation of these earthquake publics and postdisaster public information infrastructures.

The Integrated Public Alert and Warning System   In addition to the array of planned public information releases, there are other ways in which the government envisions reaching out to people who may be affected by a disaster. For example, the Integrated Public Alert and Warning System (IPAWS), an infrastructure run by the federal government, is supposed to make it easier for government officials to issue alerts to disaster survivors or people in danger. It is billed as a one-way communication system that is robust and redundant, and might actually have a shot at working after a major disaster.[37] IPAWS is the epitome of a system that envisions the government as the convener of authority, because it gives the government the power to reach everyone with scary news. This is not an unprecedented idea; it builds on the Emergency Alert System, which originated during the Cold War to allow the President to speak via the radio to all Americans in the case of a nuclear attack.[38]

IPAWS allows authorities to send emergency alerts to people in a specific geographic area via radio, television, cell phone, computer, and even signs by the side of the road. The idea behind the program is that an "alerting authority" can submit a single message about an emergency to the

IPAWS Open Platform for Emergency Networks, and that the single "alert aggregator/gateway" will validate that the person writing the message is legitimate and submit their message to the different channels available to be disseminated.[39] One of the many techniques that the IPAWS system integrates is Wireless Emergency Alerts. This allows government authorities to specifically target the mobile phones in a physical region that the government specifies, and push alerts to these phones with network priority.[40] By integrating technologies such as Wireless Emergency Alerts with other technologies such as the Emergency Alert System, the work of reaching the public through a multiplicity of means is centralized with IPAWS.

Wireless Emergency Alerts and IPAWS are examples of what disaster response strategy documents call "the system of systems approach," which means that multiple technologies and organizations can work together using standards and protocols. These are "at the center of the Bay Area's vision for emergency public information and warning."[41] Messages must be submitted in the format of an international data standard called the Common Alerting Protocol (CAP). Utilizing this standard means that a number of fields have to be filled out in specific prescriptive ways that predetermine the structure of the message as well as its potential contents. For instance, the certainty of an event has to be described as one of the following options: "observed, likely, possible, unlikely, or unknown."[42] Protocols and standards are so powerful precisely because they allow a layer of abstraction that might make a message "readable" to a variety of different machine configurations.[43] Protocols such as CAP are also limiting; they prescribe a particular format for alert messages that shape the message in particular ways that may or may not support the goals of the alert.

A FEMA training course about the IPAWS program illustrates how FEMA envisions that the IPAWS system would shape people's experiences of a disaster. The course explains how the long arm of emergency management can reach the US public and alert it to danger through virtually any media device.[44] The course video takes viewers on a tour of a town to see how an IPAWS alert reaches people who are driving to work, watching TV at home, at a game at a stadium, and at school. A white man in a suit behind a desk occasionally interjects narration with voice-over comments. An elderly couple at home in bed hears a siren and turns on their National Oceanic and Atmospheric Administration weather radio to hear a tornado warning. In the classroom, a red scrolling alert crawler is projected onto the front of the classroom, and the teacher says, "Students, the emergency siren has gone off. We will now be leaving the room in an orderly fashion." In this video, though, these visions of a technologically savvy and prepared public seem optimistic at best. The image of a classroom with an alert system crawler connected to a national alert system seems a bit far-fetched from a financial and logistical point of view. Although the IPAWS system does allow for multiple languages to be transmitted, the scenario assumes that messages will be in English, and this alert may (or may not) be translated.[45]

The government envisions itself as central source of information that includes organizations that produce information for earthquake publics using templates such as "Suggested Information Releases" and technology projects such as IPAWS. Yet both the "Suggested Information Releases" and IPAWS also point to a paradox: the government utilizes networked distribution systems based on the tenets of other open communication systems, but envisions that its voice will be authoritative.

The State and Social Media Corporations

After a disaster, the government must work within an information order that is partially of its own making, through its production of information for earthquake publics, but also participate in an information order dominated by social media companies. In the United States, social media companies mediate both how people get news and interpersonal relationships, and are influential in shaping contemporary event epistemology. Early Internet proponents imagined that it might be a platform that would make it possible for all voices to be broadcast. Social media platforms have made it easy for people to broadcast themselves; ideally they allow for a plurality of voices. This seems to be at odds with the government project of producing authoritative information, yet these projects are very much intertwined with one another.

In 1989, in the aftermath of the Loma Prieta earthquake, the government imagined that the media would transmit its message to the public. Recently, the government has been adjusting its practices to communicate with potential earthquake publics on platforms that they already use. As crisis informatics researchers Axel Bruns and Jean Burgess observe, "Over the past decade, social media have gone through a process of legitimation and official adoption, and they are now becoming embedded as part of the official communications apparatus of many commercial and public-sector organizations—in turn providing platforms like Twitter with their own source of legitimacy."[46] Social media are a key dimension of the contemporary information order. A series of research projects undertaken by Leysia Palen and her colleagues at Project EPIC have examined the evolving use of social media after crises since the mid-2000s, such as using

Facebook after shootings to identify the deceased in 2007, sharing information through Twitter following floods in 2009, and reaching out through Facebook to locate experts during floods in 2013.[47] The analyses of varied "emergent" social media practices after US disasters make it clear how central social media corporations are to organizing postdisaster information practices.

While use of social media has become more ubiquitous, governments have come to see that social media must be mastered to reach earthquake publics and influence the information order. The press release templates provided in San Francisco's emergency plan specifically instruct denizens to engage further with the governments' information products via social media.[48] Other plans say that government-issued social media should ensure the propagation of a government tweet by asking "the public to rebroadcast the content and share it with others who they know may be affected by the incident."[49]

The California Office of Emergency Services uses a number of popular social media platforms including Google, YouTube, Facebook, and Twitter to distribute information.[50] For example, the California Office of Emergency Services' YouTube channel includes videos of press conferences and even videos intended to recruit people to use new emergency alert systems like Wireless Emergency Alerts.[51] Government emergency services organizations often relied on other media organizations, such as television and newspaper companies, to create publicly digestible "news" about emergency operations work. Speaking to earthquake publics directly via social media channels is, in some ways, novel, and requiring new kinds of information work from the agencies. In other ways, it's a new platform for the same techniques; broadcasting press conferences is hardly an innovative form. However, while most municipalities have created social media accounts, it is not clear that they are actually well practiced at *using* social media.[52] Thus, while the capabilities to connect directly with citizens via social media exist, it is not necessarily practiced well by all municipalities, which might have limited resources or skills to do this work.

Different government organizations see the Internet and social media not only as a site where they can broadcast their interests but also as platforms on which particular types of behavior from earthquake publics should be encouraged. San Francisco's postdisaster sample press release template, for instance, includes instructions to "post your status on social media" in order to update friends and family.[53] Social media are bound up in the plans for San Francisco's government to communicate with the populace via Twitter, Instagram, Facebook, and YouTube.

These privately owned platforms are also imagined to organize how San Francisco's government advocates that citizens prepare for and respond to an emergency. The city and county of San Francisco's Department of Emergency Management has a web presence with sf72.org, which advises people on how they should plan for a disaster, including the equipment that they should purchase. Sf72.org is especially focused on the idea of community and promotes the notion that people get to know their neighbors.[54] There are also specific suggestions for how people should engage with particular social media platforms. The webpage for sf72.org proposes that people create Facebook groups "with your inner circle so you can easily send messages, share supplies, and make a plan together" as well as instructions for what to do after a disaster: "Post your status to Facebook to let friends and family know you are okay." Furthermore, this site advocates using Twitter to follow an emergency account run by the San Francisco emergency organizations. It also explicitly asks people to "post about yourself and your area. Use #SF72 to include them in the crowdsourced emergency feed."[55] The website asks people to upload copies of documents such as their passport, driver's license, and birth certificate to Google Drive to "access your documents remotely in case you can't get home. Know that there is a safe copy in the cloud."

While it is pragmatic and realistic for the city of San Francisco to expect that its citizens turn to social media in the time of a disaster, as I will discuss later, it is also troubling because earthquake publics are not able to hold these private companies responsible for ensuring that they equitably enhance their infrastructure to enable this kind of communication.

Social Media as a Resource for Situational Awareness

The government uses social media to make sense of a disaster. When FEMA used social media to improve situational awareness after Hurricane Sandy in 2012, social media was important not just for circulating public information but also helping the government understand the disaster. Outside the government, researchers and businesses have recognized social media's potential value as a source of information about disasters, particularly as a source contributing to perpetual situational awareness. Though now mostly defunct, products such as Geofeedia attempted to knit Twitter data into situational awareness, seeking to produce new optics for governments that purchased the software.[56] Other contemporary volunteer-driven organizations such as the Standby Task Force, Crisis Mappers, and the Digital Humanitarian Network attempt to process social media and other data sources for government and aid organizations to use in disaster response.[57]

Academia-based research projects have aimed to "leverage the public's collective intelligence" via social media data to "support situational awareness," and give people a "top-down" and "big picture" view of the disaster.[58] Papers about these projects describe the various mechanisms for processing Twitter data (including the incorporation of various networked services) in order to contribute to situational awareness. There are many different aspects of the data assemblage that go into producing this kind of situational awareness: the sources of the Twitter data (the people whose awareness is being culled), the visualization platforms that make the data sensible to humans, the massive financial, technical, and organizational infrastructure that supports the Twitter platform, and the military-academic complex that supports many of these researchers.

Whether used by the government, private researchers, or academics, the manner in which Twitter data is used to represent reality is riddled with biases and ethical issues. Researchers have shown that social media platforms and organizations that attempt to use their content for disaster response purposes reproduce the inequities in societies experiencing disasters as well as the inequities between humanitarians and people suffering after disaster.[59] Users are assumed to have consented to allow their data to be used for disaster response, and the Twitter data is always incomplete because many people—often including those most adversely affected by a disaster—are not included.[60] The type of situational awareness that Twitter can produce is shaped by access to proprietary tweets and regulated by Twitter, Inc.[61] Researchers and entities must negotiate access to the data sets by paying for them and using various application programming interfaces. Even if a company or researcher manages to have access to all Twitter data, it is difficult to identify all of the tweets relevant to a disaster, interpret the context of short tweets, and verify their veracity.[62] In order to make the tweets usable to increase situational awareness, researchers have to use other services, particularly geolocation ones. Thus, the situational awareness that is produced by Twitter frequently relies on a "messy assemblage" of other services, thereby reshaping and deforming the world it attempts to bring the analyst closer to.[63]

Lucy Suchman describes some of these in her discussion of the messy assemblage that produces situational awareness during war. In military situations, the various media used to create situational awareness enables people who operate drones remotely to believe they understand a situation enough to decide who to kill. And Suchman makes it clear that the stakes of situational awareness—"the messy assemblage of socio-technical mediation"—are high: identifying objects incorrectly can lead to accidentally killing civilians.[64]

The stakes for situational awareness in the disaster context are different; in theory, situational awareness allows decision makers and those in charge of resources to decide what to do as well as where to deploy those resources.[65] Theoretically, Twitter data sets could enable a small number of key decision makers to decide to use their resources to save some people, while others, who may not be visible on social media, perish.[66] The distortions that the messy assemblages producing situational awareness introduce are not obvious because many pieces are owned by social media companies, which are not transparent about what data they collect, what they do with it, and what portion of it is available to whom.

Commercial Media and Information Authority

In the aftermath of a disaster, the government assumes two roles: as a consumer of public information, via sociotechnical assemblages for situational awareness, and a producer of public information for citizens, the delivery of which is supported by a number of different public information infrastructures. From the perspective of producing and circulating public information, the capabilities of the state seem to have greatly expanded in recent years. New systems, such as IPAWS, offer many more avenues for many different types of government officials to target specific groups of people with public information. But the government has also made use of commercial systems, such as YouTube and Twitter. This aspect of public information infrastructure—privately owned and embedded in the information practices of the state—is one that deserves scrutiny.

Beyond the role of the government as a producer of public information, the government is a consumer of reports generated by earthquake publics. Although it was not in the interest of "situational awareness" specifically, during the 1989 earthquake the government tried to make sense of what was going on, and to a certain degree, it was unsuccessful. This was in some sense because of the practical limits of what people could do during a disaster as well as the government's imagining of the populace as English-speaking and middle class, and the state's symbiosis with the media. But importantly, in the 1989 disaster response plans, earthquake publics were not imagined as a source of understanding for the government. Today, there is an opening up of the types of legitimate voices that can contribute to the government's awareness of what has happened, such that the voices of people on social media have been incorporated into the government's plans (though this is not to say that they will inform action). At the beginning of this chapter, I presented the ideas of social media and the government's plans for producing public information as dialectical. Yet ordinary people's voices are potentially folded into the government's situational awareness, which in turn informs the production of government public information. Even though there is more room for the voices of various earthquake publics to be incorporated into the government's imagination of what is happening after a disaster, the government is still in the powerful position of deciding (or not) to listen to and legitimize certain voices, and these public information infrastructures—especially the ones including social media—have important limitations built into them.

Calculating Earthquake Publics

As the previous chapters have detailed, people want to find each other after a disaster and learn about what has happened. These wishes are persistent today, though within a different information order where there are new corporations that intermediate interpersonal relationships. And these corporations want to remain central to the public's experience of a disaster in new ways; there are products in development for use after a disaster that enroll people in a platform's information collection and centralization project.

In 2014, Mark Zuckerberg, the CEO of Facebook, launched a project called Safety Check.[67] If Facebook believed that someone was in an area that had been affected by a disaster and Facebook has decided to run the Safety Check product, this person would receive a notification from Facebook to confirm their safety. Facebook explains the process this way:

When the tool is activated after a natural disaster and if you're in the affected area, you'll receive a Facebook notification asking if you're safe.

We'll determine your location by looking at the city you have listed in your profile, your last location if you've opted in to the Nearby Friends product, and the city where you are using the internet.

If we get your location wrong, you can mark that you're outside the affected area.

If you're safe, you can select "I'm Safe" and a notification and News Feed story will be generated with your update. Your friends can also mark you as safe.[68]

The design of the Safety Check tool constructs what Tarleton Gillespie calls a "calculated public," or a public assembled through algorithms. Imagining an earthquake public through the reality described in one of the many products attempting to produce situational awareness also produces a calculated public: a group of people that is a sociotechnical creation of complex, privately owned algorithmic and data assemblages.[69] The public that is calculated by an analysis of Twitter is limited by the data that constitutes Twitter as well as the algorithms used to produce and analyze this data.

When it was first deployed, the calculated Safety Check public was primarily made up of those who Facebook believed were in harm's way and needed to update their status. Secondarily, the calculated public included "friends" who were alerted about safety statuses. The calculated public was determined by examining all users that logged on until Facebook found a user who it believed was in the area that it had specified geographically as a "disaster area."[70] Facebook looked at all the users who were connected to this user as friends and determined if any of those people were in the area, and then repeated this process with all these people's friends. This was effective because any two random users on Facebook were connected by less than five "hops."[71] Facebook tested the application with employees before making it available to its users. These tests revealed that receiving multiple requests from Safety Check asking you to mark yourself safe was actually "really stressful." The developers were also aware of the ramifications of errors within Safety Check: "The last thing you want to feel buggy is a product around safety for earthquakes and stuff like that."[72] Indeed, as previous chapters have illustrated, the emotional process of finding out about the status of loved ones is fraught; a bug in a calculated public that is supposed to represent those in an affected area could cause distress. In 2017, Facebook launched another product, called "Community Help," which offers a formal space for people to offer and request resources such as food, water, and shelter after a disaster.[73] Facebook also introduced tools that allowed some users to personally fundraise after a disaster, from which Facebook takes a processing fee, allowing Facebook to directly profit from disaster.[74]

The desire to account for the missing has also motivated other earlier humanitarian technology projects. Person Finder is a product developed by Google.org, a nonprofit funded by Google. Like Safety Check, the focus of Person Finder is on accounting for the well-being of loved ones. The Person Finder website can be deployed by Google.org after a disaster. As with technologies that support disaster relief run by Facebook, Google.org gets to choose when it deploys Person Finder, and has no explicit public metrics available for when a decision is made or what criteria it uses.[75] This system presents users with two different options: "I'm looking for someone," and "I have information about someone." If you click on the button to say that you are looking for someone, you are presented with a single search box.[76] Anyone can enter a record, and anyone can look someone up.[77] The Person Finder system was built on the People Finder Interchange Format, a public standard for describing missing people developed after Hurricane Katrina in 2005, when many (in the eyes of some) redundant websites cropped up where people could post the names of missing or found people. Google.org first built Person Finder in 2010 after the earthquake in Haiti. People can access the data by searching for the names of their loved one.[78]

Person Finder embodies the desire to make it possible to find other people by creating records about people and centralizing many information systems into one—a description that could also apply to the registration project after the 1906 earthquake. The Google.org project is also a helpful comparison in light of Safety Check: while Safety Check asks people whether they are safe, Google.org's Person Finder requires that someone actively come to the website and hence may limit usage.[79] The Google.org product, however, can share the data with disaster responders and other lifesaving organizations, and uses a public data format.

In its disaster preparedness materials, the city of San Francisco encourages people to use Facebook. This reinforces the idea that Facebook is a place where people *should* register their status, and cements the company at the center of an earthquake public's informational experience of a disaster. The centrality of these companies in mediating people's personal and emotional experiences of a disaster is similar to the role of newspapers in the 1906 earthquake and fire, when newspapers published lists of people who registered with them. Where newspapers were key to convening particular earthquake publics in 1906 or 1868, social media are pivotal to convening some calculated earthquake publics today. If people have Internet connectivity after an earthquake—and this is, of course, an *if*—products such as Safety Check will likely be used and important to people's knowledge about the earthquake. And even if people don't have connectivity, the fact that they can't use Safety Check is also likely to be important to their event epistemology. Products such as Safety Check would seem to challenge the authority of the government to manage all aspects of disaster response, yet because of the state's adoption of corporate social media platforms as sites to broadcast its messages and learn about what is going on, these social media calculated publics are fixed into the government's disaster response.

Questioning Calculated Earthquake Publics
The use of Twitter, Facebook, and other corporate platforms raises questions about who these calculated earthquake publics are, what kinds of publics

are worthy of these platforms being deployed, the companies that own the platforms that instantiate these publics, and what kinds of responsibilities these companies might bear in their capacity as disaster responders.

Both Safety Check and Person Finder raise questions about who is made visible, and who is not; while in theory everyone is included in these systems, people may have trouble using them when they don't have access to computers or the Internet. Additionally, these systems are troubling for privacy advocates. The Person Finder system makes information about people not just legible but visible as well—with all the potential advantages and perils that comes with visibility.[80] In the case of Person Finder, people can find out much faster if their family is safe, but one can also imagine a myriad of ways these systems could be misused.[81] As with most technologies, registration systems have values and intentions built into them, but these values and intentions do not determine their uses.

The ability of Facebook to accurately guess who might be affected by a disaster—to calculate the public—is questionable. In Nepal, Facebook invited almost three million people to respond to the Safety Check within five minutes of the earthquake. A number of people marked themselves as "safe" after the Nepal earthquake though they were safely located in the United States.[82] Furthermore, the utility of Safety Check is limited because its imagined user must have connectivity and technology access immediately after an earthquake, so that the user will be able to alert Facebook that they are safe. In a place where people don't use Facebook, the calculated earthquake public is limited.[83] The earthquake publics that are calculated by technology companies are necessarily limited to people who have access to technology—and this excludes those who cannot access platforms such as Facebook because they don't have the financial resources.

But it is not just the algorithmic choices of the Facebook developers that shape disaster communication; it is also the existential decision of whether the Safety Check or Person Finder product is appropriate for a particular event. In other words, is a specific earthquake public "worthy" of being a calculated earthquake public? Users of Facebook have questioned why some events have provoked Facebook to make use of Safety Check product, while other events have not warranted deployment. For example, Facebook deployed Safety Check after attackers murdered and injured hundreds in Paris on November 13, 2015, yet only one day earlier, a similarly brutal act of terror in a Beirut suburb did not prompt any such deployment. Facebook vice president of growth Alex Schultz responded to this criticism by noting that the company is still trying to figure out when it is appropriate to deploy the technology, but that with "an ongoing crisis, like war or [an] epidemic, Safety Check in its current form is not that useful for people: because there isn't a clear start or end point and, unfortunately, it's impossible to know when someone is truly 'safe.'"[84] The implication is that disasters with a clear period where lives are in danger are when a technology such as Safety Check is appropriate. Since the 1970s, however, disaster researchers have been trying to explain that simplistic notions of temporality with disasters ignore the long-term structural inequities that shape the outcomes of disaster.[85] Something like Safety Check is really only useful for determining if someone is alive, and not the implications of a disaster for person. The political issues associated with the ontological problem of disaster did not escape Facebook, and in 2017 they outsourced the problem of declaring emergencies to two other companies—iJET (an "integrated risk management" company) and NC4 (this company is an information management company trying to "reduce threats").[86]

The Facebook algorithm is able to traverse the entire "graph" of the Facebook user base relatively quickly. Of course, finding people quickly is what Facebook is ideally designed to do—for the purpose of advertising to people. Facebook has plenty of economic reasons to make sure that people think that it has developed a trustworthy system. Remaining central to people's experience of a traumatic event helps Facebook, like the newspapers of 1906, to retain its dominance. Furthermore, both Facebook and Google have the money to devote resources to make products that may be useful to people in a variety of situations. Facebook apparently did not put many "resources" into developing its Safety Check product because it was "unproven" (possibly a typical practice for all Facebook products).[87] This observation supports the conclusion that creating robust and accurate disaster response information systems is not a project that Facebook deems worthy of the effort of many engineers. There is nothing inherently wrong with this, but it does raise questions about how a profit-driven company, as part of a postdisaster public information infrastructure, shapes the experience of a disaster when its interest in public welfare is going to be secondary to sustaining the corporation. What kinds of responsibilities should be required of a company involved in disaster response? Even if it is in Facebook's interest to serve the its users in difficult times with a product such as Safety Check, Facebook can still use whatever data it collects to whatever ends are covered in its broad terms of service.

Internally, Facebook referred to the Safety Check notifications as a kind of "mass messaging," though the messages are targeted.[88] Facebook is under different regulatory regimes than companies that are considered broadcast companies by the Federal Communications Commission and thus required by law to participate in the Emergency Alert System.[89] It has no legal responsibility to help people during disasters, nor is there any pretense that Facebook would not exploit disaster survivors. Yet technology industry sycophants have declared that Facebook "is fast becoming one of the world's most important emergency response institutions."[90] It is therefore not just through the government's adoption of social media technologies to circulate public information, or the use of social media by public and private actors to gain situational awareness, that the government and these corporations become involved; it is also through social media corporations acting like "emergency response institutions"—traditionally the realm of state and civil society actors—that blurs the relationship between all these entities in our contemporary information order.

Conclusion

Today, earthquake publics are often calculated publics. They are calculated through the design of aspects of public information infrastructures—social media corporations—and adoption of these sociotechnical practices in ways that reify the limitations of these calculated publics. And it is not just social media platform companies that calculate publics. The government has a calculated public embedded in its imagined postdisaster information practices as well. In disaster response plans, the government envisions inclusive earthquake publics, with different languages and abilities, but these same plans also imagine making sense of disaster impacts by using particular technologies that are not always inclusive. The next major earthquake, as the government aims to shape it, will occur amid a "risk-conscious public."[91] In practice, for the government, this means that there are multiple means for alerting people about risks, and perhaps optimistically, that the government envisions the earthquake publics as more of a partner in interpreting risks, and disasters.

While the government's vision of the public is inclusive, its conception of itself as an information producer is thoroughly hierarchical. The government both produces and processes information through NIMS—a highly standardized and uniform organizational scheme that envisions a singular path for producing authoritative public information through the Joint Information System. At the outset of the chapter, I posited that the modes of producing and circulating information for the government and social media corporations seemed to ideally be dialectical information practices—the government uses social media, and social media companies step into the role of disaster responders.

Even as new disaster response plans try to account for these blurring lines, and while these plans and commercial social media platforms shape the future, they do not determine it; the 1989 earthquake illustrates the limitations of planning. Assessments of the preparedness of counties in the San Francisco Bay Area show that though disaster response plans may promise inclusiveness, and technologies may facilitate this, the government may still not actually be able to reach out to many groups of people after a disaster.

As the 1906 earthquake and fire aftermath demonstrated, dominant institutions shape event epistemologies. But surprising work-arounds will also likely be practiced to meet the extraordinary information needs of people facing future disasters. Of course, the past serves as a limited guide to our contemporary event epistemology when we consider the power of distant technologists to analyze earthquakes with new computational techniques and use these stories to influence on-the-ground action.

6 Comparing Information Orders: Continuity and Change

How do experiences of earthquakes shift between the various historical information orders that I explore in this book? Disaster informatics researchers have considered how social media has impacted events such as the Arab Spring uprisings in 2011 and Hurricane Sandy in 2012, and pundits gleefully declared that new infrastructures represent a new way of knowing. Yet few have ventured into examining historical disaster information practices that might validate these claims. Throughout this book, some aspects of past information orders might feel totally unfamiliar, while others are relatable. I borrowed the term *information order* from Bayly to describe the institutions, infrastructures, and practices that shape how information is produced, circulated, shared, and used as a means of surveillance and control. This book looked at Northern California information orders in 1868, 1906, 1989, and today, and asked how the information orders shaped postdisaster knowledge. I introduced the term *event epistemology* to show that information orders and disasters coconstruct possibilities for knowing. For example, an accurate count of the number of people who died in the 1906 earthquake and fire was not part of the event epistemology. This was in part because accumulating the physical records was challenging, but also because a low number of deceased was politically convenient. The destruction wrought by earthquakes creates intense interest in information about what happened at the very moment the institutions, infrastructures, and practices that create and limit knowledge are challenged. Disasters test information orders and provide interesting moments to understand how information orders shape event epistemologies.

It is here, in this conclusion, that I return to the comparative question, having laid out all the cases, to consider how different information orders shape event epistemologies, both for people experiencing a disaster and researchers like me today. I answer this question about shifts in event epistemology between information orders from three interconnected vantage points: information practices, material technologies, and institutional arrangements. From the perspective of information practices, there is a lot of similarity between the different earthquakes explored in this book. In 1868 and, we expect, today, people want to talk to their friends and families, and find out what happened. People engage in creating interesting work-arounds to maintain knowledge about the well-being of those they are close to. Yet the infrastructures that facilitate these practices are not open to all.

From the perspective of material technologies, there are obvious differences between the telegraph and Twitter. What has the shift in the materiality of technology meant for event epistemology? The affordances of technologies along with the pricing of technology access and practices associated with updating loved ones all influence the types of news that people might first receive. And private companies potentially convene publics in novel ways. Yet, I argue that the changes in practice are not as salient as the entanglement of the technical materialities with the shifting role of institutions and ideological approaches to disaster response.

Last, I turn to the question of institutional arrangements, examining both private companies involved in producing public information infrastructure and the work of governments. What is significant and different between the person in 1868 and the person today is not how someone finds out about the earthquake—be it from a telegram or Twitter; rather, it is that a multitude of public institutions exist today, each with a sophisticated and complex set of anticipatory bureaucratic processes in place to produce public information and make meaning from the information produced by others. And though the impulse to call mom after a major earthquake has not changed, the ability for private companies to potentially collate all the contacts with mom and use them to summarize the effects of an earthquake has important implications for the constellation of institutions involved in disaster response along with the types of actions that they might take. Disaster informatics must not only account for shifts in event epistemologies that are related to the affordances of new material technologies but also the shifts in institutional actors and the activities they undertake.

Earthquakes, as disasters, are unique in their unpredictability and suddenness as well as the types of physical damage they can cause. Yet the insights about information orders that I develop below are not necessarily limited to postearthquake contexts. In the United States, which has adopted an "all-hazard" approach to disasters—meaning that the response to most events that could be called disasters uses the same disaster response framework—there are parallels in the expectations for the state's response to an earthquake, destruction caused by climate change, and a violent terrorist act. Disasters also challenge the information order, and as I have shown throughout the book, what happens in the aftermath reveals the ways that power and authority are reinforced.

Information Practices

One way to approach the question of the impact of different information orders is to consider the information practices of earthquake survivors. To analyze this, I developed the analytic of *public information infrastructure* from existing scholarship on information infrastructures and the formation of publics. The empirical chapters of this book show how public information infrastructures in times of disasters cannot simply be understood as entities that are "out there"; they must be seen as continual relational processes that are always being produced—not just by the powerful companies that own the wires, or employees who fix wires, or government organizations who execute disaster response plans, but also by the earthquake publics that instantiate the public information infrastructures. The work to produce public information infrastructures can be observed in moments when infrastructure is injured by an earthquake, and the ongoing process of making and circulating information continues, albeit in an altered form.

After an earthquake, there is an intense need for information, making public information infrastructures a focus of activity. As *earthquake publics*, groups of people interested in making sense of the consequences of the disaster, wonder what has happened to loved ones and more generally what the results of the terrifying earthquake are, they report on what they know and their own statuses, and their knowledge is constituted in public information infrastructures. In all the earthquakes examined in this book, immediately after the earthquakes, people who weren't in the disaster area worried about those who had experienced the earthquake, and the people who survived the earthquake wanted to soothe the concerns of their friends and families. People made great efforts to notify others about their well-being, whether the means of doing so was by letter, telegraph, or phone (and in contemporary times, we must add the Internet to this list). Maintaining knowledge about the status of friends and loved ones was an ongoing project for many people in many of the historical moments in this book, not just an artifact of contemporary information technologies.[1]

The project of accounting for others is ongoing, but it is unusually intense after a disaster. Understanding how people become informed about their loved ones in extraordinary circumstances tells us about how event epistemologies are constructed. I explored the problem of maintaining knowledge about the status of loved ones most thoroughly in the 1906 chapter. In 1906, the San Francisco earthquake and fire caused massive destruction, and forced many people to relocate after their homes burned to the ground. In this extreme instance, people were left without any knowledge of what had happened to their friends and families. San Francisco residents such as Sarah attempted to go to multiple telegraph offices so that she could tell others that she had survived. People hung signs on fences, advertised in local papers, wrote letters that were mailed for free, and sent telegraphs in order to find out how others had fared. After the earthquake and fire, earthquake publics had to confront and adapt to the problem of accounting for people in an environment of physical destruction. In a very real sense, the people who were accounted for were represented through public information infrastructures.

The need to get in touch was intense and emotional; concern for those who were in the area of the 1906 earthquake was high, and survivors were equally concerned about their worried relations. The telegraph operators accepted Sarah's telegraph even though they were extremely backed up, and telegraphs eventually ended up being transported by train or through the postal system. While many people initially turned to the telegraph to quickly get in touch with people who were far away, it was actually the post office that was able to deliver the telegrams. As people waited for telegrams to be delivered, they imagined the worst and were enormously relieved when they were incorrect. The concern that flowed in both directions—from people who were affected and those who were worrying about what had happened to their beloveds—was intense and realized through the public information infrastructures. The processes involved in the public information infrastructures enabling people to get in touch with each other involved both innovative work-arounds and well-known practices. The public information infrastructures helped people get in touch with loved ones and eased emotional anguish following the devastation. After the spectacular damage of the 1906 earthquake and fire, the work of making the public information infrastructure function was challenging and revealed the powerful entities capable of meeting people's information needs.

Deep-pocketed institutions (such as the newspaper companies) were able to recover quickly, and because of the pressing need for news after a disaster, reinforced their centrality in the public information infrastructure. Powerful companies with extensive resources dominated the information order and even personal communications in this moment. Individuals in the earthquake publics that I described getting in touch with friends and families had to contend with the story of the earthquake that the newspaper, radio, and television companies already circulated. In the earthquakes of 1868, 1906, and 1989, the sensationalizing press was so distrusted that people who lived through the earthquakes were highly concerned that friends and families would be frantic worrying about their safety. Earthquake survivors wrote letters and made phone calls in the wake of disaster news to correct exaggerated or downplayed stories as well as soothe loved ones. Of course, damage to public information infrastructures often made communication challenging, while also being a source of news itself. Through all the earthquakes, the working and nonworking status of public information infrastructures was a way of interpreting the disaster. Earthquake publics went to extraordinary lengths to do the work of sharing their personal status, even with broken technology, thereby creating work-arounds to facilitate personal communication in the direst of times. Survivors' reported well-being was mediated through the public information infrastructure not only via the news that circulated but also by the people who interpreted the working and nonworking status of public information infrastructures to answer the question, "How bad?"

As ongoing sociotechnical processes, public information infrastructures shape the lived experiences of earthquake publics in the manner in which they produce and distribute knowledge. Yet public information infrastructures all necessarily have exclusions. Importantly, in the historic cases examined, public information infrastructures worked differently for the rich and poor, English speakers and those not fluent in English, and permanent residents and the transient. The public information infrastructures knit together certain earthquake publics while excluding others, and reifying inequities. Public information infrastructures such as the various registration systems were not inclusive. To the extent that the public information infrastructures were crucial for letting people know personal well-being, they had real impact on the emotional states of earthquake survivors and people waiting to hear from others affected by the earthquake. Nonetheless, alternative public information infrastructures emerged and facilitated these traditionally excluded populations, such as Chinese Californians, in finding each other.

Alternative public information infrastructures are crucial to understanding public information infrastructures in 1989. The event epistemology was shaped by the government's goal to prepare public information, articulated in disaster response plans and enacted by complex bureaucracies. Like in 1906, alternative public information infrastructures emerged to support the non-English-speaking publics out of the structural exclusion of non-English speakers. In 1989, radio broadcast—not conventionally thought to be a medium for personal broadcast—helped connect Spanish speakers far from the earthquake to their loved ones who survived it.

From the perspective of information practice, information orders in different eras permit and even condition people to keep track of loved ones. After an earthquake, people in all the earthquakes explored in this book wanted to hear from families and friends about what had happened, and public information infrastructures, as modern assemblages, made this possible. In fact, when public information infrastructures were injured, people made inspired efforts to figure out other ways of getting news. Even when the infrastructures that facilitated these practices were not open to all.

The Materiality of Public Information Infrastructures

Across the Northern California earthquakes examined in this book, there was stability in practices associated with maintaining knowledge about loved ones, yet there has also been a notable shift in the material dimensions of those practices. The material qualities of the technologies of the information order have changed, and so too have notification practices. Earthquake news in all these cases included a mix of personal and mass communications. For people far away, the first news about the earthquake was generally from a media company, not a personal connection. In 1868, and even in some cases in 1906, we saw that people initially received news from a media source, but then followed this up with news from a more personal source. In 1868, William Henry Knight wrote to his mother to provide his own account of the earthquake, knowing that she had already heard a sensationalized story from the media. In some sense, this was the result of the technical and social limitations of the telegraph at that time; everyone could not simultaneously send a telegraph, and telegraph company pricing decided which messages got sent—meaning it was too expensive for many people. The idea that people's personal stories followed the mainstream media's stories often meant that people needed to provide particular details about what happened, because they needed to correct a story with their own interpretation. In 1989, people wanted to deliver personal news simultaneously with the stories that the mainstream media circulated, but in many cases the overburdened phone networks did not allow this. Over all the earthquakes, people wanted to share their personal observations of what happened with their loved ones—they wanted to tell their "war story"—but this sharing usually had to wait until after the newspaper, TV, or radio stories had circulated.

In historic California earthquakes, the news of the earthquake might have reached people quickly, but the details from personal connections of what had happened might have had to wait because of the limits of what information was prioritized with the limited bandwidth in public information infrastructures. Today people learn news from journalistic institutions and from their friends and families simultaneously. The personal story is frequently the report circulated first, particularly on social media platforms, since producing well-researched news stories takes more time than personal updates. In the earthquakes in 1868, 1906, and 1989, the personal updates were delivered after media reports, and they shaped the way that news was shared because they reacted to earlier stories; similarly, today when personal accounts are shared first, they can influence the media reports that are subsequently shared. On the one hand, assemblages that enable the speedy sharing of personal accounts allow for the quick soothing of people concerned about the safety of others, and even further the broad sharing of stories that might not otherwise be represented in journalistic or state reports. But on the other hand, these technologies allow lies and deliberate misrepresentations to spread like wildfire because there is no accountability process.[2]

Over time, the materiality of public information infrastructures, working in concert with pricing and other practices, has influenced the sharing of news; the speed of knowing generally that an earthquake has occurred has not much changed, however, even for people far away. By this I mean that in 1868 with a working telegraph, people in Boston could "instantly" hear about an earthquake in California. Yet once the telegraph was sent to a telegraph office in Boston, it was not instantly news to everyone in Boston. The news would have to be broadcast in some way. In both the 1868 and 1906 earthquakes, people across the country crowded around bulletin boards or at telegraph offices to get the latest updates about what had happened. In 1989, many Americans were sitting at home watching the World Series, which was suddenly interrupted by the news of the earthquake. In these moments, there was a huge outpouring of concern about what had happened, and everyone with a television had access to the updates from the national media. Today, the pervasive availability of news is even more complete; if people have the Internet on their phones, and their phones are available, they might instantly have access to news of an earthquake. Still, the speed of the news, especially between 1989 and today, seems to be somewhat marginal. People can find out quickly what is going on, but not *so much* quicker, even than in 1868. Thus, it is not purely the technical speed of the news but also the political, social, and economic forces that organize the materials that determine which acts of distribution and inter- pretation matter postdisaster.

The materiality of public information infrastructures also shapes the kinds of publics that public information infrastructures can convene. In 1868 and 1906, letters from earthquake survivors were read aloud to inter- ested parties in public spaces or printed in newspapers. Whose letter was read or reprinted, and who heard the content of letters, was embedded in the social practices of the day. People gathered at telegraph offices to hear the latest news of an earthquake. And different newspapers, based on their design, printing equipment, and printing frequency and distribution plan, influenced who read the reprinted letters.

Similarly today, the design of technologies shapes the contours of pub- lics. Gillespie argues that new technologies convene novel formulations of publics, or what he calls "calculated publics."[3] Corporations such as Face- book convene calculated earthquake publics. The calculated earthquake publics are initially determined algorithmically and offer little visibility into how the software actually constructs these earthquake publics. Social media platforms also set the terms of how people interact with others after a disaster through design choices that ask for specific types of speech, display certain news, and promote particular interactions. Furthermore, focusing on the technical aspects of social media platforms paints a limited picture of the world that excludes nonusers.[4] And by concentrating on the techni- cal affordances of particular platforms, it can be easy overstate technology at the expense of acknowledging the labor that goes into producing the public information infrastructures as well as the institutions that dominate disaster response.[5] Social media companies remain opaque on whether they cohere with the values of fairness and freedom that government public information production processes imagine—a topic that I expand on below.

Beyond the way that the materiality of technology shapes the kinds of publics that convene, it also enables a type of social imaginary: the total archive.[6] The ability to collate many digital sources quickly, afforded by the Internet along with a myriad of storage and algorithmic technologies, allows for the imagining of a kind of specificity in what Donna Haraway calls the "God's eye."[7] New technologies have led some people to believe that they can now have a type of omniscient situational awareness even though they are located far away from where a disaster occurred. The ability to collate and automatically process many sources of news has radically changed. While one could have received telegraph messages and pinned these reports about damage to a map, this is more easily done in volume with digital technologies. Using text messages with geotagged information, or with the help of volunteers (another formation of a calculated public) adding geocoded information, people can quickly collate many reports about damage onto maps or filter results through various other analytic tools.[8]

People with the power to distribute resources use these sociotechnical platforms—as part of the public information infrastructure—to imagine that they can see the whole situation and know best how to act.[9] Drones are now being used to collect information about an area that has been affected by a disaster. Those controlling technologies must trust that these tools present an accurate picture of the disaster, so leaders can make decisions quickly about a targeted area of people.[10] As we see the technologies for making faraway places more believably knowable, we also see an expansion of the bureaucratic apparatus for responding to disasters—a phenomenon that I expand on below.

Private Companies, Publics' Information

Looking at the cases in this book, there can be little doubt that the experience of an earthquake today would be shaped by the dominant media companies. In this section, I consider the comparative information order question from the perspective of institutions, with a focus on the corporations involved in the information order. The following section examines the changing work of the state, before I finally look at the arrangement of the public and private sector in contemporary disaster response.

Private companies involved in producing public information infrastructure have had a dominant role in shaping event epistemologies throughout this book—the infrastructural technologies these companies produce have politics. They are inscribed with the logics of those who designed and built them and the organizations that they work for, and they shape possibilities in the world around them. A focus on public information infrastructures as well as its politics and affordances sometimes obscures the economic dimensions of public information infrastructures that may answer existential questions about why a particular sociotechnical assemblage exists.[11]

Throughout this book, for-profit companies are central to how many people make sense of disasters because these companies help produce event epistemologies. In 1868, the newspaper companies participated in downplaying the results of the earthquake as well as exaggerating them, and glorifying their role in circulating news. Media companies weren't monolithic in their practices or power. Some de-emphasized the earthquake to support long-term interests in San Francisco as a viable place to invest; others sensationalized the disaster to sell more newspapers. Some newspapers agitated for the production of authoritative information about the results of the disaster. The newspapers, as platforms for circulating details about the earthquake, were clearly important to the earthquake publics. Even when they weren't able to function as "normal" after the earthquake, the status of the newspapers (and functionality of the telegraph networks that supported their work) was a source of news about the earthquake's effects. The role of the newspaper companies in the information order was not totalizing in part because the companies were not trusted. Nor could the newspaper companies alone construct the event epistemology; telegrams, postcards, letters (sometimes even those published in the newspapers), and illustrations modified, supported, and contradicted the different newspaper stories.

In 1906, powerful and wealthy newspaper and telegraph companies remained central to the earthquake public's experiences of the earthquake and fire as intermediaries facilitating personal communication with loved ones. They experienced extreme damage and worked to recover quickly. And like 1868, many were involved in both downplaying and sensationalizing the earthquake damage. People who wanted to learn about the well-being of people affected by the earthquake relied on various public information infrastructures to communicate with their loved ones, and when these infrastructures broke or were overloaded, the broken infrastructure itself was a source of concern and consternation. Wealthy and powerful newspaper companies, determined to remain central to the public, were both a source of sensationalistic stories and destination for people to find evidence of the fates of companions.

The Loma Prieta earthquake is an interesting contrast to the previous ones in the sense that there was a multitude of different types of dominant news platforms: newspapers, television, and radio were all important. In 1989, the national television media's attention did not always fall on the earthquake publics that most needed aid but instead focused on spectacular imagery that captivated a national audience. The media and government reinforced each other's centrality in the disaster response: the government expected media companies to carry its messages, and the media companies frequently cited government agencies as sources. The stories that the media companies told were partial, sometimes sensational, and catered to audiences that had not actually experienced the earthquake, not unlike the previous earthquakes explored in the book. Because there were federal organizations, such as FEMA, available to provide aid, there was not the same impetus to de-emphasize the earthquake because of potential economic losses. Bad earthquake damage in 1989 actually meant *more* federal dollars for the state.

As in 1906, some of today's most highly valued companies—Facebook, Google, and Twitter—are intermediaries in public communication eager to remain central in the information order, especially in times of disturbance. In the case of earthquake publics, technology companies have important roles to play as intermediaries in both personal relationships *and* the relations between citizenry and the state, similar to the role of the media companies after the 1989 earthquake. Yet the optics that public information infrastructures produce today are unique in the arc of this book. Platform media companies collect massive amounts of data about the people who use their platforms and profit from this data by selling targeted advertising. The content on media platforms is stored, sorted, organized, and processed by people and machine learning algorithms to produce situational awareness. Processing social media data for use in disaster situations requires a large group of laborers, whose work is often overlooked in descriptions of disaster response that focus on the technological enablers of this kind of information processing.[12] Relying on social media platforms during times of crisis reifies the relations produced between the powerful platform owners and the mostly unpaid masses whose many voices make these platforms possible.[13]

Moments of disaster and great upheaval are styled by free market ideologues as clean slates, and these times of "shock" are used to advance neoliberal political agendas under the guise of the state's emergency authorities; Naomi Klein calls this "disaster capitalism."[14] Here I want to make a related though slightly different argument: the information order, when examined in different moments in time, shows that in postdisaster instances, it is not so much an issue of the *government* privatizing aspects of the infrastructure but rather the fact that wealthy, *privately owned companies* are already dominant and in a position to remain powerful because they can improvise work-arounds when things break. Corporations might even be more powerful because they can remain central to earthquake publics' experiences of disasters. Today we might say that Facebook is in a better position to connect missing families after a crisis. But this is not because Facebook is the institution that should shoulder this responsibility—the company answers to investors, not the public welfare; it is because Facebook has the resources to do this work. As this book makes clear, however, this configuration is not new. It is reminiscent of the nineteenth- and early twentieth-century earthquakes.

The Bureaucratization of Disaster Response

As crucial as companies that produce public information infrastructures are to event epistemologies, here I turn to examine the contributions of another institution: the state. Changes related to the government's role in the arrangement of institutions involved in responding to disasters are in some ways much more dramatic than those related to the material infrastructure or role of private corporations. Most notable is what I have been calling the *bureaucratization of disaster response*: that the government is given the permission to prepare for disaster response—a process that from an informational perspective, includes producing public information, and at present, ingesting and processing information produced by earthquake publics to create situational awareness. While the term *bureaucracy* has a largely negative connotation in the United States, I don't want to invite this interpretation.[15] This growth trend in government involvement in disaster response has meant that there is a party responsible for all the people affected. If citizens are left out of the response, they have some kind of recourse, at least theoretically. In 1989, for example, advocacy groups sued the government when they were left out of the earthquake response and recovery.[16]

I also don't want to give the impression that the state did not respond to disasters prior to the Cold War; the government definitely played a role in the region's responses to the earthquakes in 1868 and especially 1906.[17] In fact, ever since the Portuguese government assumed responsibility for the Lisbon earthquake in 1755, governments at some level have been involved in disaster response.[18] What is different in the post–Cold War era is the *planning* of disaster response, and for the purposes of this book, the planning of the production of public information.

Rather than react after a disaster, the state today engages in a range of "anticipatory techniques" that attempt to predict and plan for what is going to happen.[19] This process of planning is in some sense an artifact of "reflexive modernity" and the "risk society."[20] It is a way of asserting control and mastery over a risky and uncertain set of possibilities, including earthquakes.[21] The plans are also crucially available to earthquake publics and researchers like me—a contrast to the social media companies, which can choose both whether or not they want to make postdisaster plans available, or be responsible at all for helping people. One result of the anticipatory governance of state information practices is that after a disaster in the United States, earthquake publics expect that the government will be prepared to issue "public information" about what has happened and what to do next. In this book, I have focused on the complex bureaucratic apparatus set up to produce and process information, but anticipatory technologies are not limited to this area.

The next section examines the different ways that the various actors in the postearthquake information orders produce information authority. I then discuss the authority of the state in moments of emergency. Last, I look at the distribution of power among state and private institutions. As the state encourages earthquake publics to make use of social media platforms to broadcast their whereabouts, the state also uses social media and Internet platforms to broadcast public information and create situational awareness, while at the same time social media corporations create their own responses to disasters.

Struggles for Authority

The information order has always been a site of contention over which accounts of an event are treated authoritatively. In moments when people were desperate for news of the fates of their loved ones and details about what had happened, the companies and government entities that controlled the means of producing and circulating stories were in especially powerful positions. In some of the cases in this book, the entities recognized that these were moments where they could be dominant in the information order by deploying their extensive resources to quickly fix the physical public information infrastructure, such that it would again be working. Other entities attempted to dominate the information order by fueling sensational stories that captured the public's attention or manipulating the stories to meet their own agendas.

In 1868, elite businesspeople, concerned about how the earthquake might appear to people who were far away from what had happened, telegraphed a story "back East" that downplayed the damage. Because of the cost of sending telegrams as well as limits on the number of telegrams that could be sent, these businesspeople believed that their story might have legs and that they could dominate the information order. While this effort to de-emphasize the earthquake was not universally adopted, and in some cases was roundly dismissed, it helped set the terms for how others described the earthquake. These terms, the estimation of earthquake damage conveyed as the dollar amount of damage, condensed the earthquake story into a single number. Moreover, because California was not widely understood to be a site of destructive earthquakes, how the earthquake damage and recovery was represented nationally and internationally was also a referendum on the viability of San Francisco as an investment destination. Deliberations by the city government about how to rebuild ignored the seismic provisions recommended by architects. Informational authority was exercised using the contours and limitations of public information infrastructure, although it was not definitively resolved. As the 1868 earthquake illustrates, after an earthquake there is a public eager for information about what has happened, a press wanting to provide information about it, and powerful people desiring to shape the event epistemology.

While money and power remain important in all information orders, the sites of the production of information authority shift over time. Within the course of the postearthquake moments in Northern California, the 1989 earthquake shows a different constellation of powerful institutions exercising their informational authority than occurred during the 1868 earthquake; in particular, the state's role in the production and circulation of information becomes much more prominent. Within the 1989 information order, the government imagined in advance of the earthquake that it would be powerful in asserting its vision of what happened, though it failed to foresee its own limitations in producing knowledge about the earthquake as well as its relationship with commercial media. The national television networks' coverage of the earthquake was critiqued because of its stress on the most sensational damage, especially the wealthy parts of the Bay Area. The government in 1989 envisioned that it would produce authoritative public information about the earthquake for all survivors, but its own response was not inclusive of non-English speakers and was very much shaped by media companies' coverage of the earthquake that centered on visually spectacular damage.

In addition to seeing the shift in how anticipatory governance techniques shaped disaster response, there was a change in the type of sources available to me as a researcher that also signal a shift in who claimed authority to talk about the earthquake. The records available about the 1868 earthquake are almost entirely newspapers, with a few letters, photographs, and other illustrations to augment them. As I discussed in that chapter, many efforts to make formal reports failed. On the other hand, in 1906, my sources include not just newspapers but also military, scientific, and Red Cross reports. The military and Red Cross were active participants in the response as well as in defining what happened after the earthquake, vigorously explaining their perspectives on the earthquake and the work that they did. After the 1989 Loma Prieta earthquake, government organizations such as the California Office of Emergency Services and FEMA claimed the authority to explain and respond to the earthquake, and wrote reports about what had happened. A cousin to the idea of bureaucratic rationality is the idea that modern institutions produce documents about themselves. Ulrich Beck and Anthony Giddens both formulate a version of this—one that is less focused on documents—as reflexive modernity. In Beck's *Risk Society*, modern institutions produce and use documents to help construct, manage, and apprehend risk. These reflexive institutions shape information orders in both the moment and retrospectively. In both 1906 and 1989, the institutions participating in the disaster response produced documents about their own activities.

Government disaster plans are also changing. Disaster response plans, as public information infrastructure, aim to create authoritative information about disasters through complex organizational structures and omniscient sociotechnical assemblages. Attention to the presence of multiple alternative public information infrastructures prompts questions about who the government is serving. The earthquake public, as an active producer of "information" as well as a population containing nonwhite, non-middle-class, non-English speakers, is acknowledged in a way in the current plans that it was not in the 1989 ones. Recent government disaster response plans imagine earthquake publics to be more inclusive of the most vulnerable populations—at least in the plans. Yet assessments of preparedness note that many municipalities in the Bay Area are not prepared to assist with multilingual populations.[22] Furthermore, government plans clearly acknowledge the public as a potential producer of useful public information via government monitoring and use of social media.

If this book only looked at the much-celebrated changes in communication technologies after disasters, we would miss the important rise of the government as the major force shaping disaster response and the focus on producing public information that controls the narrative. This bureaucratization of disaster response is arguably the most significant development in postearthquake information orders explored here and key for thinking about the future of disaster informatics because it shapes other information and communication practices. The bureaucratization of disaster response asks analysts of information practices to account for and question the crucial role of state institutions in shaping event epistemologies.

Emergency Power

The expansion of the state's techniques to produce authoritative information raises questions about the consequences of these institutional configurations. Examining historic information production practices after disasters reveals different configurations of the state and private sector in producing public information infrastructures. In the chapter on the 1989 earthquake,

I showed how fraught the idea of producing a planned, inclusive authoritative account of a disaster is. Yet this book suggests that regardless of how difficult it is for the government to attempt to produce public information about an earthquake that is usable to all, earthquake publics still see this as government's responsibility. Some may argue that this production of public information is a "foisted benefit" of government that the citizenry doesn't ask for and the government should let people individually figure out how to meet their own information needs. The case studies in this book indicate otherwise. Contrasting the earthquakes of 1868 and 1989 is instructive here.

The 1868 earthquake serves as an example of when the role of the government in general, and federal government in particular, was minimal in producing public information. An intense interest by the public in accurate accounts of what had happened did not mean that a "correct" (to borrow the terminology of the capitalists trying to downplay the 1868 earthquake) narrative about the disaster emerged. Powerful entities—such as the business class—produced self-serving narratives. For the business class, this narrative meant minimizing the earthquake damage because it was bad for business in the San Francisco Bay Area. For other entities, though, such as some newspaper companies and other printers, the narrative was the opposite: sensationalizing the damage gave them the opportunity to sell more postcards, newspapers, and letterheads. The local governments still had their own self-interest in portraying the disaster in a specific way. But this particular set of relations did not yield accounts of the disaster that resulted in increased building safety standards, better science, or more truthful accounts of the earthquake's impacts despite calls to make changes. In later years, government intervention after the earthquake of 1989 on behalf of the earthquake publics had a positive impact on the ability of the San Francisco Bay Area region to cope with an earthquake.[23] I found no evidence of earthquake publics complaining of too much involvement on the government's part after the earthquake; only that it failed to serve all earthquake publics.

Disasters are a domain in which the government might extend its powers in ways that are otherwise distasteful; because people are desperate for help, the government is granted emergency authority to spend money and deploy resources. In disastrous moments where people are desperate, we have generally agreed as a society that the government has the right to deploy resources without going through the usual channels and be granted emergency authority for collecting information.[24] The government adopting novel information collection practices in times of crisis is problematic because of the legacy of emergency measures becoming permanent law.

For example, the PATRIOT Act was supposed to be an emergency measure after the September 11, 2001, terrorist attacks. Civil liberties groups have long decried the problems of the surveillance enabled by the PATRIOT Act, including a lack of transparency along with a decrease in citizens' autonomy.[25] Many of these rules are nonetheless still on the books today. Extraordinary practices introduced in emergency times became normalized and permanent. Yet when people consent to surveillance in a desperate moment, without understanding the full ramifications of such an action, they are giving their consent under duress; this consent has no moral force. Further, it is difficult to prove that this incursion into informational norms produces a tangibly different quality of governance response because it means engaging in counterfactual speculation.[26] Emergencies have been moments for the expansion of state power, and this includes its informational power.

But not all government power requires emergency authority. As Stephen Collier and Andrew Lakoff argue, state power can be exercised through anticipatory techniques.[27] Disaster planning is an anticipatory technique that does not require emergency legal powers. Governments publish disaster plans to demonstrate mastery over uncertainty; they apply risk management techniques, which require both complete information about the system as well as anticipation of the future.[28] The project of mitigating risk is key to the modern risk society. According to analysts of this concept, the risk society is one where in our attempts to order the world, we produce the conditions of destruction that are distributed unevenly throughout our society.[29] According to this thinking, the public information infrastructures that we rely on are also complex, with uncertain points of failure in disaster moments, and should be planned and managed.[30] Furthermore, anticipatory disaster techniques are often adopted from warfare planning; since the advent of US disaster response planning, there has been a tension around how civilian disaster response is influenced by military or, since 9/11, security logics.[31] When the state extends anticipatory governance technologies into the "everyday" domain, airport security, for instance, concerns arise about the securitization of daily life.[32] Utilizing anticipatory governance techniques can invoke a sense of permanent crisis as well as extend many of the concerns about the trampling of information privacy norms in emergency situations into the everyday.

From Bureaucratization to Privatization
At the same time that I examine the state's role in producing and distributing informational authority, I want to ask the same questions of the media companies as institutional actors in disaster response. As this book noted in

the chapter on the present-day information order, although the state and private companies may appear to represent distinct interests, in the information order their interests are intertwined.

Today the government not only seeks to produce public information but also collects information from social media platforms. On the one hand, it seems important that the state tries to understand the experiences of various earthquake publics. Monitoring social media is a crucial way for the government to learn about what is happening on the ground. Authorities can identify those who need help quickly and get the aid to those desiring it. On the other hand, this kind of surveillance can enable control from far away, leading to a lack of local autonomy. Moreover, when the state incorporates privately owned social media platforms into its public information practices, it also implicitly assumes the logics of the private companies that generate particular formations of companies' calculated publics. This enfolding of the calculated logics of capitalist enterprise within the state response is not only potentially problematic because of the infusion of economic goals into welfare-oriented programs but also because it threatens to undo the types of inclusive earthquake publics that the state hopes to produce.

The state has not just assumed the capitalist logics of the social media companies; these corporations attempt to adopt powers traditionally associated with the state. The contemporary platform owners have taken the initiative to roll out particular products when they decide that something is a disaster. This feels problematic to people, because declaring an emergency is indicative of sovereignty.[33] In other words, it is uncomfortable when private entities such as Facebook declare emergencies because it seems like the act of a state entity. Declaring emergencies is a powerful ontological act. We are uncomfortable with even governments using it because other countries have entered into permanent states of crisis where governments get permanent emergency powers. The fear with private companies is similar. When a private company decides that it needs to, in some sense, change the rules and declare an emergency, such as running a "Safety Check" of all the people that it decides are potentially affected by a disaster, this feels like an action that a government might take.

Speculating one step further, if people hold their cell phones as embodiments of themselves, products like Safety Check reach into people's private spaces in unauthorized and alarming ways. People might find the idea of giving private companies responsibility for the well-being of a society abhorrent because it requires a level of trust in the companies that it is not clear they deserve. Or people might see that there are enough incentives in place for these companies to behave according to the wishes of the earthquake publics. While citizens theoretically have visibility into when a government can use emergency powers, we do not understand the limits of emergency with private companies.

Continuity within Information Orders

This book asks a grand, comparative question about the information orders in different historical and contemporary events. But it also asks about how the public information infrastructures recover, or don't, after each earthquake. Examining public information infrastructures in existential moments (to use the extreme language of extreme events) helps us understand the stabilizing actions that do or do not keep infrastructures going. In the last section of this chapter, I return to the question of what happens when the public information infrastructures that people so desperately need to work cannot function as they do on an everyday basis.

Stepping back from noting the consistencies and changes in event epistemologies between the different information orders, I want to consider how the public information infrastructure after each earthquake did or did not recover. After all the crises that I explored, earthquake publics were anxious to learn about the welfare of friends and families, but also about the scope of the damage. The world around them had suddenly changed, and they didn't know how. One can see this desperation for news reflected in the media diaries of the San Jose University students after the 1989 Loma Prieta earthquake as well as in the self-congratulatory details of the newspaper companies quickly selling out of their postearthquake issues in 1868. Overall, though, the companies involved in producing public information infrastructures had adequate resources, and were able to recover quickly, retain their status in the information order, and alleviate the desperation for news felt by earthquake publics. Though there were genuinely innovative work-arounds that arose in the moments when public information infrastructure broke, they were mostly aimed toward making things work as they did before the earthquake. The means and modes for circulating the details of what happened become extremely important, and form the backbone of my earlier argument that the earthquake publics are in part instantiated through these public information infrastructures.

It is well known that in disasters, it is the most vulnerable who usually suffer the most. In these historical moments, the reverse of this was also true, those with excess resources generally fared well. The earthquakes were an opportunity for those institutions in the public information infrastructures with money to exert power because they could use their surplus resources to recover quickly. The fates of newspaper companies after the earthquakes portray a nimbleness and willingness to use interesting work-arounds to print their papers regardless of whether the letters used for type-setting were strewn across the floor (as in 1868), presses were destroyed (as in 1906), or there was no electricity (as in 1989). For organizations involved in the public information infrastructure, these moments after a disaster were opportunities to be centrally important to the earthquake public by producing stories at the instant that people were hungriest for news of an updated version of the postearthquake world.

Stephen Graham and Nigel Thrift contend that today we overempha-size catastrophic collapse, and in the process underemphasize the kind of mundane repair efforts that actually make infrastructures function.[34] Even in the parts of San Francisco that were completely destroyed in 1906, the ways that people found each other were often rooted in information and communication practices from before the earthquake. Stories about the operability of the telegraph, post office, and registration bureaus implicitly suggest a narrative of continuity. When the physical infrastructure broke, work practices and institutions maintaining infrastructure endured. The way that people fixed the public information infrastructure frequently reflected predisaster methods and practices for gathering as well as sharing information. The earthquake publics' event epistemology after the 1906 earthquake and fire was not only shaped by the physical attributes of the infrastructure but also by the various publics, media companies, telegraph businesses, post offices, government organizations, and relief charities that participated in continually producing the public information infrastructure in the face of extreme difficulty.

Scholars of information infrastructures have long noted their durability in light of new designs and planned changes. Here I extend this finding to the disaster context. For the earthquakes I analyze in this book, the domi-nant institutions tended to stay dominant in times of disaster, as people came up with interesting work-arounds to enable the continuity of public information infrastructures. The technologies and practices that people use regularly are also the technologies and practices that they lean on in times of disaster. Certainly there are exceptions to this: some earthquake publics needed to create alternate public information infrastructures or registration systems for their information needs. These exceptions, however, almost always have a foot in the predisaster information order. While we may be tempted to label these exceptions as innovative to satisfy discursive trends, we should resist that temptation and instead acknowledge the inspiring challenge of continuity.

Notes

Preface

1. This remembrance and others below it are quoted in Patrick Sheriff, ed., *2:46: Aftershocks: Stories from the Japan Earthquake* (London: Enhanced Editions Ltd., 2011).

2. James D. Goltz and Dennis S. Mileti, "Public Response to a Catastrophic Southern California Earthquake: A Sociological Perspective," *Earthquake Spectra* 27, no. 2 (2011): 494.

3. Someone who calls themselves Our Man in Abiko, his partner, and a group of volunteers that Our Man in Abiko had never met before all assembled the book. They donated all the profits from the book to Japan's Red Cross. Sheriff, *2:46*.

4. Because of the particular network of people who produced *2:46*, the story is probably heavily shaped by the experiences of people from the United States or Canada, or people who know people from the United States or Canada.

5. Abiko resident Miho Nishihiro opined, "I think the biggest problem has been in the transfer of information. The only thing I can praise the Tokyo Electric Power Company for is the rapidity and accuracy of reporting radiation levels to the public." Sheriff, *2:46*.

Chapter 1

1. C. A. Bayly, *Empire and Information: Intelligence Gathering and Social Communication in India, 1780–1870* (Cambridge: Cambridge University Press, 1996).

2. For an example of a related argument about the power of discourse in disasters, see Kathleen Tierney, Christine Bevc, and Erica Kuligowski, "Metaphors Matter: Disaster Myths, Media Frames, and Their Consequences in Hurricane Katrina," *Annals of the American Academy of Political and Social Science* 604, no. 1 (March 2006): 57–81.

3. Bayly, *Empire and Information*, 366.

4. Ibid., 366–367.

5. "What we have called the information order should not be seen as a 'thing,' any more than a state or an economy is a thing; it is a heuristic device, or a field of investigation, which can be used to probe the organisation, value and limitations of past societies. It is not separate from the world of power or economic exploitation, but stands both prior to it and dependent on it. It can be considered to have a degree of autonomy from politics or economic structure." Ibid., 4.

6. In the science and technology studies tradition, infrastructure is "a broad category referring to *pervasive enabling resources in the networked form.*" Geoffrey C. Bowker, Karen Baker, Florence Millerand, and David Ribes, "Toward Information Infrastructure Studies: Ways of Knowing in a Networked Environment," in *International Handbook of Internet Research*, ed. Jeremy Hunsinger, Lisbeth Klastrup, and Matthew Allen (Dordrecht: Springer Netherlands, 2010), 98.

If infrastructures are considered enabling resources until they are not enabling and break, the visibility of infrastructures are often thought to be central to their definition: "The sudden absence of infrastructural flow creates visibility, just as the continued, normalized use of infrastructures creates a deep taken-for-grantedness and invisibility." Stephen Graham and Nigel Thrift, "Out of Order: Understanding Repair and Maintenance," *Theory, Culture, and Society* 24, no. 3 (May 1, 2007): 8.

Through his studies of water in Mumbai, Nikil Anand shows how the visibility of infrastructure is sometimes a necessity for those continually involved in the process of making infrastructure work for them when it fails to bring them water. Nikhil Anand, "PRESSURE: The PoliTechnics of Water Supply in Mumbai," *Cultural Anthropology* 26, no. 4 (November 2011): 542–564; Nikhil Anand, "Municipal Disconnect: On Abject Water and Its Urban Infrastructures," *Ethnography* 13, no. 4 (December 1, 2012): 487–509; Nikhil Anand, "Leaky States: Water Audits, Ignorance, and the Politics of Infrastructure," *Public Culture* 27, no. 2 (January 1, 2015): 305–330.

That is, unless the point is producing invisibility. Olga Kuchinskaya takes on the idea of visibility by specifically interrogating "the production of invisibility … the practices of producing representations that limit public visibility of Chernobyl radiation and its health effects." As Kuchinskaya notes, the effects of Chernobyl radiation are basically invisible, and understanding the effects is a fundamentally mediated process. The visibility of the infrastructure that produces evidence of radiation is experienced only by a few. In the work of Anand, Kuchinskaya, and others, the ability to control the visibility and invisibility of infrastructure is a source of privilege and power. Olga Kuchinskaya, *The Politics of Invisibility: Public Knowledge about Radiation Health Effects after Chernobyl* (Cambridge, MA: MIT Press, 2014).

7. Bowker et al., "Toward Information Infrastructure Studies," 98. Christine Borgman succinctly summarized Susan Leigh Star and Karen Ruhleder's original description of the dimensions of information infrastructure,

Their eight dimensions can be paraphrased as follows: An infrastructure is embedded in other structures, social arrangements, and technologies. It is transparent, in that it invisibly supports tasks. Its reach or scope may be spatial or temporal, in that it reaches beyond a single event or a single site of practice. Infrastructure is learned as part of membership of an organization or group. It is linked with conventions of practice of day-to-day work. Infrastructure is the embodiment of standards, so that other tools and infrastructures can interconnect in a standardized way. It builds upon an installed base, inheriting both strengths and limitations from that base. And infrastructure becomes visible upon breakdown, in that we are most aware of it when it fails to work.

Christine L. Borgman, "The Invisible Library: Paradox of the Global Information Infrastructure," *Library Trends* 51, no. 4 (2003): 654; Susan Leigh Star and Karen Ruhleder, "Steps toward an Ecology of Infrastructure: Design and Access for Large Information Spaces," *Information Systems Research* 7, no. 1 (1996): 111–134.

8. Star and Ruhleder, "Steps toward an Ecology of Infrastructure." Christian Sandvig, "The Internet as an Infrastructure," in *The Oxford Handbook of Internet Studies*, ed. William H. Dutton (Oxford: Oxford University Press, 2013), 86–108.

9. Ibid; Paul N. Edwards, "Y2k: Millennial Reflections on Computers as Infrastructure," *History and Technology* 15, no. 1–2 (October 1998): 7–29.

10. "Infrastructures are not, in any positivist sense, simply 'out there.' The act of defining an infrastructure is a categorizing moment. Taken thoughtfully, it comprises a cultural analytic that highlights the epistemological and political commitments involved in selecting what one sees as infrastructural (and thus causal) and what one leaves out." Brian Larkin, "The Politics and Poetics of Infrastructure," *Annual Review of Anthropology* 42, no. 1 (October 21, 2013): 330.

11. Canonical works in information infrastructure studies include Star and Ruhleder, "Steps toward an Ecology of Infrastructure"; Geoffrey C. Bowker and Susan Leigh Star, *Sorting Things Out: Classification and Its Consequences* (Cambridge, MA: MIT Press, 1999); Geoffrey C. Bowker, *Science on the Run: Information Management and Industrial Geophysics at Schlumberger, 1920–1940* (Cambridge, MA: MIT Press, 1994). More recent works explicitly describing the contours of the field include Bowker et al., "Toward Information Infrastructure Studies"; Steven J. Jackson, Paul N. Edwards, Geoffrey C. Bowker, and Cory P. Knobel, "Understanding Infrastructure: History, Heuristics, and Cyberinfrastructure Policy," *First Monday* 12, no. 6 (2007): 1–9.

12. Dan Schiller, *How to Think about Information* (Urbana: University of Illinois Press, 2007); Ronald E. Day, *The Modern Invention of Information: Discourse, History, and Power* (Carbondale: Southern Illinois University Press, 2001).

13. John Seely Brown and Paul Duguid, *The Social Life of Information* (Cambridge, MA: Harvard Business School Press, 2000); Geoffrey Nunberg, "Farewell to the Information Age," in *The Future of the Book*, ed. Geoffrey Nunberg (Berkeley: University of California Press, 1996), 103–138.

14. Siva Vaidhyanathan, "Afterword: Critical Information Studies," *Cultural Studies* 20, no. 2–3 (2006): 297. Although document sounds a rather quixotic way to refer to a narrow conceptualization of information, it has a history with documentalists in the library sciences. Michael K. Buckland, "Information as Thing," *Journal of the American Society of Information Science* 42, no. 5 (1991): 351–360. For a short discussion of this word, see Michael K. Buckland, "What Is a 'Document'?," *Journal of the American Society for Information Science* 48, no. 9 (1997): 804–809; Niels Windfeld Lund, "Document Theory," *Annual Review of Information Science and Technology* 43, no. 1 (2009): 1–55.

15. Acknowledging the materiality of information dovetails nicely with recent work in history, communication and media studies, library sciences, and anthropology that seeks to establish the environmental history and impact of computing, physical nature of bits and the Internet infrastructure, and "realness" of the online world. Jean-François Blanchette, "A Material History of Bits," *Journal of the American Society for Information Science and Technology* 62, no. 6 (2011): 1042–1057; Nathan Ensmenger, "Computers as Ethical Artifacts," *IEEE Annals of the History of Computing* 29, no. 3 (July 2007): 88–87; Tom Boellstorff, "For Whom the Ontology Turns: Theorizing the Digital Real," *Current Anthropology* 57, no. 4 (August 2016): 387–407; Lisa D. Parks and Nicole Starosielski, *Signal Traffic: Critical Studies of Media Infrastructures* (Urbana: University of Illinois Press, 2015); Kirsten A. Foot, Pablo J. Boczkowski, and Tarleton Gillespie, *Media Technologies: Essays on Communication, Materiality, and Society* (Cambridge, MA: MIT Press, 2014).

16. Brown and Duguid, *Social Life of Information*; Nunberg, "Farewell to the Information Age."

17. E.g., Helena Karasti and Karen S. Baker, "Infrastructuring for the Long-Term: Ecological Information Management," in *Proceedings of the 37th Annual Hawaii International Conference on System Sciences: 2004* (IEEE, 2004).

18. This book provides a number of comparative cases where I can test this emphasis on continuity. Earthquakes can be huge disruptions. As far as the public information infrastructure is concerned, earthquakes can destroy physical infrastructure and, paradoxically, be extraordinarily helpful for people who want to make sense of what has happened. This echoes information infrastructure research that examines the development of new infrastructures, leading to a number of tensions "as practices, organizations, norms, expectations, and individual biographies and career trajectories bend—*or don't*—to accommodate, take advantage of, and in some cases simply survive the new possibilities and challenges posed by infrastructure." Paul N. Edwards, Steven J. Jackson, Geoffrey C. Bowker, and Cory P. Knobel, "Understanding Infrastructure: Dynamics, Tensions, and Design" (final report of the History and Theory of Infrastructure: Lessons for New Scientific Cyberinfrastructures workshop, Ann Arbor, MI, 2007), https://deepblue.lib.umich.edu/handle/2027.42/49353.

Another way to think about infrastructural endurance is through the idea that information infrastructures are embedded in long-lasting institutions. The quality of infrastructural embeddedness is ever present in descriptions of information infrastructure by the science and technology studies community. Studies of the emergence of global climate knowledge infrastructure, such as Paul Edwards's work, make it clear that institutions shaping information infrastructure are not just the organizations involved, such as the International Meteorological Organization, but also the governments of countries where meteorologic observations occur. Paul N. Edwards, "Meteorology as Infrastructural Globalism," *Osiris* 21, no. 1 (2006): 229–250; Paul N. Edwards, *A Vast Machine: Computer Models, Climate Data, and the Politics of Global Warming* (Cambridge, MA: MIT Press, 2010).

19. Star and Ruhleder set forth a number of dimensions of information infrastructure, as noted earlier. Many of these qualities underscore the durability of infrastructure. For example, infrastructure theorists say that infrastructure is *built on an installed base*, which limits the potential to change the information infrastructure, and that it is the *embodiment of standards*, which can take decades to develop and become widely used. The idea that infrastructure *links with conventions of practice* or is *learned as part of membership* also indicates the long relationship that people have with practicing infrastructure in order to make it resource-like. Star and Ruhleder, "Steps toward an Ecology of Infrastructure."

20. Ashley Carse, reflecting on the continual production of the Panama Canal and the environments that support it, says,

[Infrastructures] require human communities to maintain them, even as they shape those (and other) communities. Without maintenance, infrastructures crack, rust, and crumble and the political projects, promises, and aspirations that they carried dissipate as formerly connected places are disconnected. Maintenance is necessary because infrastructures—even global infrastructures—are both networked and embedded in the environments they cross and transform.

Ashley Carse, *Beyond the Big Ditch: Politics, Ecology, and Infrastructure at the Panama Canal* (Cambridge, MA: MIT Press, 2014), 219.

21. Paul Edwards proposes the word "infrastructuration" (after Anthony Giddens's "structuration") to describe the ways that people simultaneously make and are limited by infrastructure. Edwards, *A Vast Machine*. This is similar to thinking in another area that I take inspiration from: a practice theory framework that posits people act within structural constraints, and the constraints are relational and made in everyday practice. See, for example, Jean Lave, *Apprenticeship in Critical Ethnographic Practice* (Chicago: University of Chicago Press, 2011); Sherry B. Ortner, "Theory in Anthropology since the Sixties," *Comparative Studies in Society and History* 26, no. 1 (1984): 126–166. I take Edwards's idea to include a sense that physical-material artifacts constrain as well, where Lave and Ortner focus on social practice.

22. Building on a long tradition of analyses of organizational improvisation after disaster, recent research about improvisation, as it relates to emergency responses

that are planned, outlines three different kinds of organizational improvisations: reproductive, adaptive, and creative. Reproductive improvisations occur when plans can't be carried out, but when what is planned can be reproduced in another form. Adaptive improvisation takes place when the plans do not anticipate the situation and thus needed to be adapted. Creative improvisation is when the plans did not exist before the event. Tricia Wachtendorf, "Improvising 9/11: Organizational Improvisation following the World Trade Center Disaster" (PhD diss., University of Delaware, 2004); Tricia Wachtendorf and James M. Kendra, "Improvising Disaster in the City of Jazz: Organizational Response to Hurricane Katrina," in *Understanding Katrina: Perspectives from the Social Sciences* (Social Science Research Council, 2006), http://understandingkatrina.ssrc.org/Wachtendorf_Kendra.

This specific analysis of improvisation calls into question how much can be planned for after a disaster; what people do before a disaster may not be *planned* to occur after a disaster but instead would be totally appropriate for them to take on as a task because it involves much of their predisaster daily routine.

23. Star and Ruhleder, "Steps toward an Ecology of Infrastructure," 5.

24. Richard R. John, *Spreading the News: The American Postal System from Franklin to Morse* (Cambridge, MA: Harvard University Press, 1995); David M. Henkin, *The Postal Age: The Emergence of Modern Communications in Nineteenth-Century America* (Chicago: University of Chicago Press, 2006).

25. Star says that "this may be either spatial or temporal—infrastructure has reach beyond a single event or one-site practice." Susan Leigh Star, "The Ethnography of Infrastructure," *American Behavioral Scientist* 43, no. 3 (November 1, 1999): 381. Another explanation of reach asserts that information infrastructure has "a certain kind of reach over time, space, and a range of human and institutional activities." Jackson et al., "Understanding Infrastructure."

26. Janaki Srinivasan, Megan Finn, and Morgan Ames, "Information Determinism: The Consequences of the Faith in Information," *Information Society* 33, no. 1 (2017): 13–22.

27. For traditional science and technology studies of information infrastructure, the reach of infrastructure has helped scientists to share or even collaborate with scientists in other places or institutions. For example, see Bruno Latour's discussion of immutable mobiles: Bruno Latour, "Visualisation and Cognition: Drawing Things Together," in *Knowledge and Society Studies in the Sociology of Culture Past and Present*, vol. 6, ed. H. Kuklick (Greenwich, CT: JAI Press, 1986), 1–40.

28. Gareth Davies, "The Emergence of a National Politics of Disaster, 1865–1900," *Journal of Policy History* 26, no. 3 (July 2014): 305–326.

29. The term *public* has been debated and theorized for generations of scholars, and the question of what publics actually are in our contemporary era of digitized, networked resources is debated. Jürgen Habermas, in his classic text, both theorized and

described the European public sphere as a space of rational deliberation that arose during the seventeenth century. The Habermassian public sphere, as a descriptive project, is understood to be problematic because the public sphere that Habermas depicts is only inclusive of men who were white and Protestant. While Habermas's portrayal of historical public spheres is problematic, however, the normative ideal, which he posits, has been influential. For Habermas, print was pivotal to the functioning of the public sphere. Scholars of print and the Internet have often gravitated toward thinking of the public sphere through what kinds of publics might be coconstituted with technology because of the primacy of print within the Habermassian ideal. Jürgen Habermas, *The Structural Transformation of the Public Sphere: An Inquiry into a Category of Bourgeois Society*, trans. Thomas Burger (Cambridge, MA: MIT Press, 1991); Craig J. Calhoun, *Habermas and the Public Sphere* (Cambridge, MA: MIT Press, 1992).

30. Michael Warner, "Publics and Counterpublics," *Public Culture* 14, no. 1 (2002): 49–90; Elizabeth L. Eisenstein, *The Printing Press as an Agent of Change: Communications and Transformations in Early-Modern Europe*, vols. 1–2 (New York: Cambridge University Press, 1979); Benedict Anderson, *Imagined Communities: Reflections on the Origin and Spread of Nationalism*, rev. ed. (London: Verso Books, 2006).

31. Christopher M. Kelty, "Geeks, Social Imaginaries, and Recursive Publics," *Cultural Anthropology* 20, no. 2 (2005): 185–214; Christopher M. Kelty, *Two Bits: The Cultural Significance of Free Software* (Durham, NC: Duke University Press, 2008), 28; Tarleton Gillespie, "The Relevance of Algorithms," in *Media Technologies: Essays on Communication, Materiality, and Society*, ed. Tarleton Gillespie, Pablo J. Boczkowski, and Kirsten A. Foot (Cambridge, MA: MIT Press, 2014), 167–194.

32. Noortje Marres, *Material Participation: Technology, the Environment and Everyday Publics* (Houndmills, UK: Palgrave Macmillan, 2012).

33. Lave and Wenger write about their triptych "legitimate peripheral participation" as "a descriptor of social practice that entails learning as an integral constituent," and specify that "each of its aspects is indispensable in defining the others and cannot be considered in isolation. Its constituents contribute inseparable aspects whose combinations create a landscape—shapes, degrees, textures—of community membership." Following Lave and Wenger, I want to put forth public information infrastructure as an integrated phrase, although as I make clear, it builds on the idea of information infrastructure. And while the idea of public infrastructure, as opposed to "private infrastructure," is a sensitizing concept within the field of geography, the public that I refer to here is in the tradition of political theory. Jean Lave and Etienne Wenger, *Situated Learning: Legitimate Peripheral Participation* (New York: Cambridge University Press, 1991), 35.

34. Here I am using public to indicate a group of people, which is nominally inclusive, coming together for political purposes, or what is called an *issue public*. Scholars such as Chris Le Dantec, Marres, and Mike Ananny have elaborated on Dewey's

notion of public in the context of modern information technology. Mike Ananny, "Networked News Time: How Slow—or Fast—Do Publics Need News to Be?," *Digital Journalism* 4, no. 4 (May 18, 2016): 414–431; Mike Ananny, "Press-Public Collaboration as Infrastructure: Tracing News Organizations and Programming Publics in Application Programming Interfaces," *American Behavioral Scientist* 57, no. 5 (May 1, 2013): 623–642; Chris Le Dantec, *Designing Publics* (Cambridge, MA: MIT Press, 2016); Noortje Marres, "Issues Spark a Public into Being: A Key but Often Forgotten Point of the Lippmann-Dewey Debate," in *Making Things Public: Atmospheres of Democracy,* ed. Bruno Latour and Peter Weibel (Cambridge, MA: MIT Press, 2005), 208–217; Noortje Marres, "The Issues Deserve More Credit: Pragmatist Contributions to the Study of Public Involvement in Controversy," *Social Studies of Science* 37, no. 5 (October 1, 2007): 759–780; Marres, *Material Participation*; Noortje Marres, "Why Political Ontology Must Be Experimentalized: On Ecoshowhomes as Devices of Participation," *Social Studies of Science* 43, no. 3 (2013): 417–443.

35. See, for example, Didier Fassin and Mariella Pandolfi, eds., *Contemporary States of Emergency: The Politics of Military and Humanitarian Interventions* (Cambridge, MA: MIT Press, 2010); Craig J. Calhoun, "The Idea of Emergency: Humanitarian Action and Global (Dis)Order," in *Contemporary States of Emergency: The Politics of Military and Humanitarian Interventions,* ed. Didier Fassin and Mariella Pandolfi (Cambridge, MA: MIT Press, 2010), 29–58; Michael Guggenheim, "Introduction: Disasters as Politics—Politics as Disasters," *Sociological Review* 62 (June 1, 2014): 1–16; Manuel Tironi, "Atmospheres of Indagation: Disasters and the Politics of Excessiveness," *Sociological Review* 62 (June 1, 2014): 114–134; Naomi Klein, *The Shock Doctrine: The Rise of Disaster Capitalism* (New York: Metropolitan Books, 2007); Kathleen Tierney, *The Social Roots of Risk: Producing Disasters, Promoting Resilience* (Stanford, CA: Stanford University Press, 2014); Gareth Davies, "Dealing with Disaster: The Politics of Catastrophe in the United States, 1789–1861," *American Nineteenth Century History* 14, no. 1 (March 2013): 53–72; Davies, "Emergence of a National Politics of Disaster"; Ted Steinberg, *Acts of God: The Unnatural History of Natural Disaster in America,* 2nd ed. (New York: Oxford University Press, 2006); Michael Watts, "On the Poverty of Theory: Natural Hazards Research in Context," in *Interpretations of Calamity from the Viewpoint of Human Ecology,* ed. Kenneth Hewitt (Boston: Allen and Unwin Inc., 1983).

36. "The most important departure from the hazard/event/behavior focus that had characterized the field since the 1950s was the refinement of the concept of vulnerability, which looks at those aspects of society that reduce or exacerbate the impact of a hazard." Anthony Oliver-Smith, "Theorizing Disasters: Nature, Power, and Culture," in *Catastrophe and Culture: The Anthropology of Disaster,* ed. Susanna M. Hoffman and Anthony Oliver-Smith (Santa Fe, NM: School of American Research Press, 2002), 27.

37. Klinenberg effectively demonstrates that studying a disaster purely as a physical phenomenon may not offer explanations for the number of deaths experienced. Eric

Klinenberg, *Heat Wave: A Social Autopsy of Disaster in Chicago* (Chicago: University of Chicago Press, 2002), 230–231, 91.

38. Kim Fortun, *Advocacy after Bhopal: Environmentalism, Disaster, New Global Orders* (Chicago: University of Chicago Press, 2001); Watts, "On the Poverty of Theory."

39. Just as working infrastructure is not experienced evenly by everyone, broken infrastructure is also uneven: "Indeed, in many cities around the world, instead of addressing the needs of vulnerable groups and communities, major state and corporate investments go to sustaining continuities of flow and circulation, by providing backups to key electrical, communication, data processing, transport, or other infrastructures that sustain the main nodes and enclaves of the globalized corporate economy." Furthermore, Stephen Graham asserts, "the political nature of infrastructure disruption is often rendered invisible by media discussions of such events as 'technical' malfunctions or environmental 'Acts of God.'" Stephen Graham, "When Infrastructures Fail," in *Disrupted Cities: When Infrastructure Fails*, ed. Stephen Graham (New York: Routledge, 2009), 15–16. On broken infrastructure, see also Steven J. Jackson, "Rethinking Repair," in *Media Technologies: Essays on Communication, Materiality, and Society*, ed. Tarleton Gillespie, Pablo J. Boczkowski, and Kirsten A. Foot (Cambridge, MA: MIT Press, 2013), 221–240.

40. Simon Marvin and Stephen Graham, *Splintering Urbanism: Networked Infrastructures, Technological Mobilities and the Urban Condition* (London: Routledge, 2001).

41. Susan Leigh Star, "Orphans of Infrastructure: A New Point of Departure" (paper presented at the Future of Computing: A Vision, Oxford Internet Institute, University of Oxford, March 2007); Geoffrey C. Bowker, Stephan Timmermans, Adele E. Clark, and Ellen Balka, eds., *Boundary Objects and Beyond: Working with Leigh Star* (Cambridge, MA: MIT Press, 2015); Marvin and Graham, *Splintering Urbanism.*

42. Tierney, *Social Roots of Risk*, 147.

43. Mirca Madianou, "Digital Inequality and Second-Order Disasters: Social Media in the Typhoon Haiyan Recovery," *Social Media + Society* 1, no. 2 (2015).

44. The inclusivity of Jürgen Habermas's public sphere has been extensively critiqued. See, for example, Calhoun, *Habermas and the Public Sphere*; Nancy Fraser, "Rethinking the Public Sphere: A Contribution to the Critique of Actually Existing Democracy," *Social Text* 25–26 (1990): 56–80; Warner, "Publics and Counterpublics."

45. Disaster researchers, after assessing disaster studies literature, proposed that:

disaster research has shown that once those who have experienced a sudden onset disaster have ensured their own survival and assisted others in their immediate vicinity, they typically seek information. In ambiguous situations, the information sought may involve a credible definition of what just happened and whether the danger has passed. They certainly want to know the status of family and friends who may have been affected by the event and report their own status to those outside the area of impact. In situations in which the disaster agent is readily identifiable, information sought may include an understanding of how extensive the impact was, how officials are responding and how to secure assistance. In short, people seek to define the situation. Goltz and Mileti, "Public Response to a Catastrophic Southern California Earthquake," 494.

46. Moments of irregularity in infrastructure are key opportunities to investigate infrastructure: "Studying moments when infrastructures cease to work as they normally do is perhaps the most powerful way of really penetrating and problematizing those very normalities of flow and circulations to an extent where they can be subjected to critical scrutiny." Graham, "When Infrastructures Fail," 3.

Information infrastructure researchers have frequently taken the idea that infrastructure is invisible until it breaks down as a core tenant (see note 6 in this chapter for a discussion). Star and Ruhleder, "Steps toward an Ecology of Infrastructure." Yet Brian Larkin notes, "In urban theory, infrastructures are often noted for their invisibility, their taken-for-grantedness, until they break down or something goes awry; but in the colonial and postcolonial context, infrastructures command a powerful presence, and their breakdown only makes them more visible, calling into being governments' failed promises to their people as specters that haunt contemporary collapse." Brian Larkin, *Signal and Noise: Media, Infrastructure, and Urban Culture in Nigeria* (Durham, NC: Duke University Press, 2008), 245.

Sometimes the builders of infrastructure want people to appreciate and even revel in it. "The fetishization of networks dwells exactly in the twin condition that connection to the network implies acquiring the use value of the utility *and* realizes the promise of participating [in] the phantasmagoric new world of technological advancement and 'progress'; a world in which human freedom and emancipation resides in connection to the technological networks." Maria Kaika and Erik Swyngedouw, "Fetishizing the Modern City: The Phantasmagoria of Urban Technological Networks," *International Journal of Urban and Regional Research* 24, no. 1 (2000): 124.

For Kaika, Swyngedouw, and Larkin as well as David Nye, who describes the "American Technological Sublime" as "a defining ideal, helping to bind together a multicultural society," infrastructures help societies define themselves. David E. Nye, *American Technological Sublime* (Cambridge, MA: MIT Press, 1994).

47. As Larkin observes, infrastructure has "poetics" and material qualities. Larkin explains how sometimes infrastructures are viewed as quintessentially modern: "Roads and railways are not just technical objects ... but also operate on the level of fantasy and desire. They encode the dreams of individuals and societies and are the vehicles whereby those fantasies are transmitted and made emotionally real." Larkin, "Politics and Poetics of Infrastructure," 133.

48. "Infrastructures are largely responsible for the sense of stability of life in the developed world, the feeling that things work, and will go on working, without the need for thought or action on the part of users beyond paying the monthly bills. This stability has many dimensions, most of them directly related to the specific

nature of modernity." Paul N. Edwards, "Infrastructure and Modernity: Force, Time, and Social Organization in the History of Sociotechnical Systems," in *Modernity and Technology*, ed. Thomas J. Misa, Philip Brey, and Andrew Feenberg (Cambridge, MA: MIT Press, 2003), 188.

49. One particular form of governance has been called "vital systems security." Stephen J. Collier and Andrew Lakoff, "Vital Systems Security: Reflexive Biopolitics and the Government of Emergency," *Theory, Culture, and Society* 32, no. 2 (March 1, 2015): 19–51.

50. Modern people and institutions are at once trying to order nature (in the case of this book, that includes producing infrastructure), but at the same time crises themselves are also central to modern life and ordering the working of society. Kevin Rozario, *The Culture of Calamity: Disaster and the Making of Modern America* (Chicago: University of Chicago Press, 2007), 23.

Rozario builds on the work of sociologist Ulrich Beck, who says that a new "risk society" is shaped by new risks created by modern technology, and how people distribute or manage risk. Ulrich Beck, *Risk Society: Towards a New Modernity*, trans. Mark Ritter (London: Sage Publications, 1992). Jean-Baptiste Fressoz has traced the concerns over distributing risk back to the nineteenth century. Jean-Baptiste Fressoz, "Beck Back in the 19th Century: Towards a Genealogy of Risk Society," *History and Technology* 23, no. 4 (2007): 333–350.

51. Joseph Schumpeter said that this "process of creative destruction" outlined by Karl Marx shows that the "evolutionary character" of capitalism is not simply due to war or population increases but is also intrinsic to the nature of capital accumulation as well as the search for innovation and new sites for capital investment. For some, Schumpeter's destruction is not a negative necessity but instead an opportunity for advancement and rapid change. Joseph A. Schumpeter, *Capitalism, Socialism, and Democracy*, 3rd ed. (New York: Harper and Row, 1976), 82–83.

Some economists have considered creative destruction in the context of disasters. These authors summarize other research on creative destruction and disaster, and dismiss it, as they find no evidence that disasters improve developing economies using "knowledge transfer" proxies such as importation of new technology. Jesús Crespo Cuaresma, Jaroslava Hlouskova, and Michael Obersteiner, "Natural Disasters as Creative Destruction? Evidence from Developing Countries," *Economic Inquiry* 46, no. 2 (2008): 214–226.

The many stories of post–Hurricane Katrina New Orleans illustrate problems with the "social imaginary" of the catastrophic logic of modernity: the disaster was followed by the displacement of hundreds of thousands of people along with the privatization of municipal and other state-funded institutions. Charles Taylor, *Modern Social Imaginaries* (Durham, NC: Duke University Press, 2004); Klein, *Shock Doctrine*; Graham Owen, "After the Flood: Disaster Capitalism and the Symbolic Restructuring of Intellectual Space," *Culture and Organization* 17, no. 2 (March 2011): 123–137; Rozario, *Culture of Calamity*.

52. Klein, *Shock Doctrine*.

53. David M. Levy, "Fixed or Fluid?: Document Stability and New Media," in *Proceedings of the 1994 ACM European Conference on Hypermedia Technology* (New York: ACM, 1994), 24–31.

54. Researchers who combine historical and qualitative social science approaches inspired me, such as Dianne Vaughan's "historical ethnography" and Mary Des Chene's "archive as field site." Diane Vaughan, *The Challenger Launch Decision: Risky Technology, Culture, and Deviance at NASA* (Chicago: University of Chicago Press, 1996); Mary Des Chene, "Locating the Past," in *Anthropological Locations: Boundaries and Grounds of a Field Science*, ed. Akhil Gupta and James Ferguson (Berkeley: University of California Press, 1997), 66–85.

55. As David Ribes and Charlotte Lee assert, "Breakdowns themselves are a kind of natural infrastructural inversion." David Ribes and Charlotte P. Lee, "Sociotechnical Studies of Cyberinfrastructure and E-Research: Current Themes and Future Trajectories," *Computer Supported Cooperative Work* 19, no. 3–4 (August 2010): 238.

56. Vaughan, *Challenger Launch Decision*.

57. See ibid., appendix C. Vaughan says that "ethnographers must consider what went unrecorded, what documents are missing, and what the effect of this historic sifting and sorting is upon the record available to us. The construction of the surviving documentary record also must always be questioned. ... Ethnographers reconstructing history must be wary of how these same factors bias their own selection process." Diane Vaughan, "Theorizing Disaster: Analogy, Historical Ethnography, and the Challenger Accident," *Ethnography* 5, no. 3 (September 1, 2004): 338. I also adapted some of Vaughan's techniques in assembling an event chronologically, especially in the chapter on the 1868 earthquake, to understand how the event unfolded in newspapers to people experiencing it.

58. Edwards, *Vast Machine*, 432.

59. Star, "Ethnography of Infrastructure," 388.

60. It is challenging to keep up with the ongoing developments of various technologies and disaster response organizations. The research for this chapter mostly concluded by the end 2016, and the findings reflect that era.

Chapter 2

1. Knight was the manager of San Francisco's successful Bancroft Publishing Department, the famed publishers of California history after California became a state. William Henry Knight to his mother, October 25, 1868, 1906 San Francisco Earthquake and Fire Digital Collection, Bancroft Library, University of California at Berkeley, 1–3, http://ark.cdlib.org/ark:/13030/hb6r29p1h5; John Walton Caughey,

"Hubert Howe Bancroft, Historian of Western America," *American Historical Review* 50, no. 3 (April 1945): 461.

2. The US Geological Survey report notes that "property loss was extensive … and 30 people were killed. The total property loss was about $350,000." The Lawson report said that the "total list of casualties due directly to the earthquake numbered 5, and about 25 more occurred from secondary causes." Carl W. Stover and Jerry L. Coffman, *Seismicity of the United States, 1568–1989 (Revised)* (Washington, DC: US Geological Survey, 1993), 104; Andrew Lawson et al., *The California Earthquake of April 18, 1906: Report of the State Earthquake Investigation Commission in Two Volumes and Atlas* (Washington, DC: Carnegie Institution of Washington, 1908), 439.

3. Thomas M. Brocher et al., *The Hayward Fault—Is It Due for a Repeat of the Powerful 1868 Earthquake?*, Fact Sheet 2008–3019 (Washington, DC: US Geological Survey, 2008).

4. Stephen Tobriner completed a careful analysis of building damage from the newspaper reports after 1868, made a map that summarized the damage, and showed that most of it was indeed, as reported in 1868, on what was called "made land." Stephen Tobriner, *Bracing for Disaster: Earthquake-Resistant Architecture and Engineering, 1838–1933* (Berkeley, CA: Heyday Books, 2006).

5. Most of the chapter relies primarily on a close analysis of several California newspapers, including California's first daily, the *Daily Alta California*, in San Francisco; the year-old *San Francisco Daily Morning Chronicle*, a newspaper that pioneered a number of "sensational" twentieth-century newspaper techniques in San Francisco; and the *Sacramento Daily Union*. Other newspapers and periodicals from Alameda County (including Oakland and San Leandro), San Jose, and San Francisco were also used. I followed the conversation about the earthquake in all these papers until the end of 1868, transcribing relevant articles. I tracked the stories about the San Francisco Chamber of Commerce's Earthquake Committee in the *Sacramento Daily Union*, *Daily Alta California*, and *Daily Morning Chronicle* until the results were presented to the chamber in 1870. Other than newspapers, I also made use of letters, imagery, and documents. I found letters about the earthquake at the San Francisco Public Library, California Historical Society, Society of California Pioneers, and University of California at Berkeley's Bancroft Library, using the available finding aids along with assistance from many helpful librarians and archivists. The California Historical Society had some useful documents related to the background of San Francisco's Chamber of Commerce that I didn't find reference to elsewhere, such as *One Hundred Years of Service: San Francisco Chamber of Commerce, 1850–1950*. The California Academy of Sciences librarians were kind enough to let me work with the precious *Meeting Minutes*; most of the other records of the academy were destroyed in the 1906 Earthquake and Fire.

I was surprised and grateful to find a number of visual resources related to the earthquake. In the process of digitizing thousands of documents and photographs related to the 1906 earthquake, the Bancroft Library and other participating organizations digitized much of the material they had available about the 1868 earthquake. Photographs of the damage from the 1868 earthquake are available at the Online Archive of California. The Society of California Pioneers has a number of resources related to letter sheets. Its copies of the illustrated newspaper editions have been invaluable. Tobriner's *Bracing for Disaster* has a number of the photographs from the Bancroft Library's collection, with his markups and analysis.

6. There are huge limitations to the sources used. In some cases, the limitations are obvious; for instance, it seems that many of the records of the San Francisco Chamber of Commerce's Earthquake Committee are lost. Unfortunately, other limitations are harder to even know. While newspapers make mention of, for example, German-speaking newspapers, I was not able to locate these non-English sources.

7. Scott Gabriel Knowles, *The Disaster Experts: Mastering Risk in Modern America* (Philadelphia: University of Pennsylvania Press, 2011); Carl-Henry Geschwind, *California Earthquakes: Science, Risk, and the Politics of Hazard Mitigation* (Baltimore: Johns Hopkins University Press, 2001).

8. Charles Wollenberg, "Life on the Seismic Frontier: The Great San Francisco Earthquake (of 1868)," *California History* 71, no. 4 (Winter 1992–1993): 494–509.

Wollenberg argues that real estate prices are paramount to understanding how San Franciscans reported the earthquake. Elite San Franciscans aimed to persuade the public that a "commercial city could prosper in earthquake country" because "the damage from the 1868 tremor was due to human error, which could easily be corrected." Ibid., 502. As Gary Brechin writes in *Imperial City*, San Francisco was built on the profits of companies that benefited from the extraction of goods from the earth. Gray Brechin, *Imperial San Francisco: Urban Power, Earthly Ruin*, 2nd ed. (Berkeley: University of California Press, 2006). See also Richard A. Walker, "California's Golden Road to Riches: Natural Resources and Regional Capitalism, 1848–1940," *Annals of the Association of American Geographers* 91, no. 1 (2001): 167–199.

9. Wollenberg asserts that despite this interest in improving antiseismic building practices, the "most ambitious attempt to study and learn from the 1868 earthquake," an earthquake committee headed by George Gordon, a real estate investor, working for the chamber of commerce, "had come to nothing." Wollenberg, "Life on the Seismic Frontier," 507.

Philip Fradkin depicts the production of the chamber of commerce report in even darker terms, observing, "Nothing was learned from the earlier experience" and thus Californians adopted a "policy of assumed indifference" to earthquakes. Philip Fradkin, *Magnitude 8: Earthquakes and Life along the San Andreas Fault* (Berkeley: University of California Press, 1999), 76, 80–81.

10. Tobriner argues against the assertion that Californians had learned nothing from the 1868 earthquake: "Historical records show that architects, engineers, and even everyday citizens understood the consequences of the earthquakes of the 1860s

and tried to inventory the damage, to understand what had happened, to retrofit buildings to resist future earthquakes, and to build earthquake-resistant structures." Tobriner, *Bracing for Disaster*, 35.

11. Olga Kuchinskaya, *The Politics of Invisibility*: Public Knowledge about Radiation Health Effects after Chernobyl (Cambridge, MA: MIT Press, 2014).

12. *San Francisco Evening Bulletin*, October 22, 1868. It seems that on the day of the earthquake, there was a paper issued in the morning at the normal time for the *Daily Morning Chronicle*, another edition issued at around 1:00 p.m., and a third edition issued at 3:30 p.m. The *Chronicle* reported that nine thousand copies of the extra on the day of the earthquake were sold. Of course this might have been pure exaggeration, but the reports of demand for news indicate the importance of the newspaper for the public. "Unparalleled Journalism," *San Francisco Daily Morning Chronicle*, October 22, 1868.

13. *San Francisco Morning Call*, October 22, 1868.

14. "Unparalleled Journalism." In that same issue of the *Chronicle*, the paper described its reporting practices: "From eight o'clock in the morning … half a dozen regular and special locals … were actively employed in gathering the facts and details of the catastrophe. Before noon all the intelligence relating to the earthquake had been collected, digested and printed. … One [local reporter] was dispatched across the bay to gather intelligence in regard to the results of the catastrophe in Alameda county."

15. One paper printed a story that the injury caused to its printing office made it impossible to print news about the earthquake itself. "By the Vallejo Route," *Sacramento Daily Union*, October 27, 1868, quoting the *Pacheco Gazette*, October 24, 1868.

16. "Brief Items," *Sacramento Daily Union*, October 27, 1868; "Laboring under Difficulties," *San Francisco Daily Morning Chronicle*, October 23, 1868; "The Earthquake in San Francisco," *Sacramento Daily Union*, October 27, 1868.

17. "The 'Bulletin' Yesterday," *San Francisco Evening Bulletin*, October 22, 1868.

18. Both the *Daily Alta California* and *San Francisco Daily Morning Chronicle* reported that the *Hebrew* newspaper was damaged badly. "Our Great Calamity," *San Francisco Daily Morning Chronicle*, October 22, 1868; "Local Intelligence: The Great Earthquake of 1868," *Daily Alta California*, October 22, 1868. Printers in the small towns south of San Francisco, such as San Jose, were reported to have an especially difficult time after the earthquake. *San Jose Mercury News*, October 22, 1868; *San Jose Weekly Argus*, October 24, 1868; "Quick Dispatch," *Sacramento Daily Union*, October 24, 1868.

19. "The Great Earthquake," *Daily Alta California*, October 23, 1868, quoting the *Oakland Daily News*, October 21, 1868.

20. The October 24, 1868, edition of the *Alameda County Gazette* opens with, "We present you, dear reader, with our earthquake edition"—its first paper issued after the earthquake on October 21, 1868. The advertising pages for the inside of the paper were only "slightly pied. ... [B]y ... judiciously patching up the dead matter which was not 'pied,'" the pages were "enabled to put in an appearance." (*Pied* here refers to the type being in disarray.) The newspaper explained that the first and fourth pages were printed before the earthquake, but that these "two principle [*sic*] advertising pages" were "worthless" so the paper would publish a "small sheet" next week.

21. The *Oakland Daily Transcript* vociferously objected to the *San Francisco Bulletin*'s accusation that its type was mixed up. Further, the *Daily Transcript* demonstrated some camaraderie and sympathy in its comments about the earthquake's impacts on the *Alameda County Gazette*: "This is rather hard upon our brother but we are satisfied that he will come out all right yet." "A Very Horny Dilemma," *Oakland Daily Transcript*, October 23, 1868.

22. *Daily Alta California*, October 22, 1868, 2. The five people listed as deceased were "William Best killed in the yard of the Occidental Hotel," W. Strong, "James B. Mansfield died on Clay street after a cornice or fire-wall falling on him," "On Taylor street, above Sutter, — —, whilst working in the back yard, was killed by a falling chimney," and "Late in the afternoon the corpse of a Chinaman, frightfully disfigured, was dug out ... on Clay street." "Local Intelligence: The Great Earthquake of 1868; List of Casualties," *Daily Alta California*, October 22, 1868.

The next day, the *Daily Alta California* reported four deaths. The casualties included Best, Strong, Mansfield, and "Kung Yung, aged forty-one, killed at or near No. 410 Clay street." "Local Intelligence: After the Earthquake," *Daily Alta California*, October 23, 1868.

23. The death of a city employee at the courthouse in San Leandro was added to the four deaths from San Francisco and "five were thus killed." *Golden Era*, October 24, 1868.

24. "General Remarks," *Alameda County Gazette*, October 24, 1868; "The Effect on Men and Animals," *Daily Alta California*, October 22, 1868; "Extra! Earthquake," *San Francisco Daily Morning Chronicle*, October 21, 1868.

25. "Pacific Slope Intelligence," *Daily Alta California*, October 25, 1868. The same report also appeared in "Effects of the Earthquake at San Francisco," *San Francisco Evening Bulletin*, October 24, 1868, quoting the *Marysville Daily Appeal*, October 23, 1868.

26. *San Francisco Evening Bulletin*, October 21, 1868.

27. "Local Intelligence: The Great Earthquake of 1868"; "City Incidents," *San Francisco Daily Morning Chronicle*, October 21, 1868; "The Earthquake in Stockton," *Sacramento Daily Union*, October 23, 1868; "Earthquake over the Bay," *San Francisco Daily*

Morning Chronicle, October 22, 1868; "Local Intelligence" and "Our Institutions of Learning," *Oakland Daily Transcript*, October 26, 1868.

28. "Earthquake over the Bay"; "Local Intelligence," *San Francisco Daily Morning Chronicle*, October 22, 1868.

29. "The scene that then ensued beggars all description—men, women and children rushed frantically forth *en dishabille* into the streets, panic-stricken, and scarce knowing whither to turn for safety, and tremblingly awaiting what the next moment should bring forth." "Earthquake over the Bay." See also *San Jose Weekly Argus*, October 24, 1868.

30. "Local Intelligence: The Great Earthquake of 1868."
Engaging in this type of reporting was typical; newspapers were not supposed to be sites of sober "objective" journalism. Michael Schudson, "The Objectivity Norm in American Journalism," *Journalism* 2, no. 2 (August 1, 2001): 149–170; David T. Z. Mindich, *Just the Facts: How "Objectivity" Came to Define American Journalism* (New York: NYU Press, 2000).

31. "Earthquake over the Bay."

32. "The Earthquake: Oakland Misrepresented," *Oakland Daily News*, October 22, 1868.

33. "Our Oakland friends would fain to have us believe that there has been no earthquake shock of any consequence in the Terminal Metropolis. This is disingenuous and absurd." "Oakland and the Earthquake," *San Francisco Daily Morning Chronicle*, October 24, 1868.

34. Other papers in Oakland directly took on the *Daily Morning Chronicle* for false reports: "Yesterday being a fine day, a large number of persons came over from San Francisco to [Oakland] who exclaimed, 'I don't see any traces of the earthquake. ... There were no trees torn up after all.' Evidently they had been reading the *Chronicle*, and expected to see things here in a much worse state than they really are." *Oakland Daily Transcript*, October 26, 1868.

35. *Oakland Daily News*, October 22, 1868.

36. *Alameda County Gazette*, October 31, 1868.

37. *San Jose Weekly Argus*, October 31, 1868.

38. All the newspapers promised that their telegraphic dispatches were in some way unique to their newspaper. The *Daily Morning Call* claimed in gothic font and bold headlines that its reports were "by Western Union Line" and "specially to this *Daily Morning Call*." The *Daily Morning Chronicle* reported that its dispatches were "special." And the *Daily Alta California* proudly declared that the telegraphic reports it had were from "the State Line."

39. Nearly identical reports in the *Call*, *Chronicle*, and *Alta* under the headline of "Telegraph" appeared from Healdsburg (written as "Heraldsburg" by the *Alta*), Redwood, San Jose, San Juan, San Mateo, San Rafael, Santa Clara, Santa Cruz, and Santa Rosa. At least two of the newspapers had the same reports for Centerville, Gilroy, Sonoma, Sonora, and Woodland. Between the three newspapers, there were only approximately a half dozen unique telegraphic reports, and only reports about Marysville were dramatically different in the different newspapers. Whatever telegrams got through on the day of the earthquake became the story that all subsequent descriptions of the earthquake must have had to acknowledge.

40. Wollenberg, "Life on the Seismic Frontier."

41. John R. Bruce, *Gaudy Century: The Story of San Francisco's Hundred Years of Robust Journalism* (New York: Random House, 1948).

The *Daily Morning Chronicle* also seemed the most willing to deviate from the story that the business elites espoused, be it in making a newspaper with illustrations of the most damaged areas or printing damage estimates before the chamber of commerce came out with its numbers. The *Daily Alta California* was the elder statesperson of the San Francisco newspapers and printed government notices. Of the newspapers that I examined closely, the *Sacramento Daily Union* presented the most distinctive story of the earthquake, perhaps because of its distance from the San Francisco elites.

42. Richard R. John, *Network Nation: Inventing American Telecommunications* (Cambridge, MA: Harvard University Press, 2010), 98; William Frank Zornow, "Jeptha H. Wade in California: Beginning the Transcontinental Telegraph," *California Historical Society Quarterly* 29, no. 4 (1950): 345–356.

43. "The Work Is Consummated," *Sacramento Daily Union*, October 25, 1861.

44. Robert J. Chandler, "The California News-Telegraph Monopoly, 1860–1870," *Southern California Quarterly* 58, no. 4 (1976): 459–484.

45. Richard B. Kielbowicz, *News in the Mail: The Press, Post Office, and Public Information, 1700–1860s* (New York: Greenwood Press, 1989), 174.

46. Richard Schwarzlose includes the *San Francisco Morning Call* in the list of Associated Press newspapers because Simonton, the owner of the *Bulletin*, who also had a financial stake in the *New York Times*, had a controlling stake in the *Call*. Richard Schwarzlose, *The Nation's Newsbrokers*, vol. 2, *The Rush to Institution, from 1865 to 1920* (Evanston, IL: Northwestern University Press, 1990); John Denton Carter, "The San Francisco Bulletin, 1855–1865: A Study in the Beginnings of Pacific Coast Journalism" (PhD diss., University of California at Berkeley, 1941), 243.

This cartel had a decade-long monopoly on eastern news because of its relationship with the New York Associated Press, which dominated the Western Union wires. The California Associated Press' domination of the eastern United States news was only challenged once the cross-continental railroad was complete in 1869 and a new telegraph company's wires were erected along this route. Chandler, "California News-Telegraph Monopoly."

During the 1860s, Western Union had a cozy relationship with the Associated Press, and gave members of this association a favorable rate. Schwarzlose, *Nation's Newsbrokers.*

Apparently Western Union raised its prices for the non–Associated Press newspaper, the *Herald*, from 6.92¢ to 15.38¢ per word. Meanwhile, the California Associated Press newspapers (the *Bulletin*, *Alta*, *New York Times*, and *Union*) went from 2.04¢ to 1.028¢ per word. "This is a conspiracy against a conservative paper to-day. Tomorrow it will be against a Republican paper, if the California Associated Press desire and decree." "The Postal Telegraph System," Daily Herald Postal Telegraph Series, document no. 1 (San Francisco: Herald Publishing Company, August 1869), 4.

47. Ears accustomed to the Morse code could circumvent monopolies over telegrams from the East. Bruce, *Gaudy Century*, 132.

48. Still, in 1869, Henry George mounted an attack on Western Union for giving newspaper companies in the Associated Press favorable rates. Henry George, letter to the executive board of the Western Union Telegraph Company, "The Western Union Telegraph Company and the California Press," April 21, 1869, Bancroft Library, University of California at Berkeley.

49. "City Intelligence at San Francisco," *Sacramento Daily Union*, October 22, 1868.

50. "The Earthquake in Sacramento," *San Francisco Daily Morning Chronicle*, October 22, 1868, quoting the *Sacramento Record*, October 22, 1868.

51. "Mrs. Smith All Right," *Daily Alta California*, October 25, 1868.

52. "The Great Earthquake: Marysville," *Daily Alta California*, October 23, 1868, quoting the *Marysville Daily Appeal*, October 22, 1868.

53. Richard R. John, *Spreading the News*: The American Postal System from Franklin to Morse (Cambridge, MA: Harvard University Press, 1995), 112–169; David M. Henkin, *The Postal Age*: The Emergence of Modern Communications in Nineteenth-Century America (Chicago: University of Chicago Press, 2006), 119–146.

54. Unable to get in touch with San Francisco, the *San Jose Daily Patriot* got a message from Santa Clara with this startling bit of news about sixty bodies. *San Jose Daily Patriot*, October 21, 1868.

55. *Daily National* (Grass Valley, Nevada County, CA), October 21, 1868.

56. "The Great Earthquake," *Daily Alta California*, October 23, 1868, quoting the *Territorial Enterprise*, October 22, 1868.

57. Ibid.

58. Ibid.

59. *Oakland Daily News*, October 24, 1868.

60. Editorial Notes, *Daily Alta California*, October 24, 1868.

61. Megan Finn, "Information Infrastructure and Descriptions of the 1857 Fort Tejon Earthquake," *Information and Culture: A Journal of History* 48, no. 2 (May 11, 2013): 194–221.

62. Gregory J. Downey, *Telegraph Messenger Boys: Labor, Technology, and Geography, 1850–1950* (New York: Routledge, 2002).

63. *Oakland Daily News—Extra!*, October 21, 1868.

64. "Weekly Trade Summary," *Daily Alta California*, October 22, 1868.

65. "The Earthquake Yesterday," *Oakland Daily Transcript*, October 22, 1868.

66. "In the afternoon the *Chronicle* issued a sensation extra, in which the disaster appears to have been greatly exaggerated; and as we do not find a corresponding statement in the [*Evening*] *Bulletin*, which appeared several hours later." "The Calamity in San Francisco," *Oakland Daily Transcript*, October 22, 1868.

67. Wollenberg, "Life on the Seismic Frontier," 502.

68. "Truth the Best Policy," *Sacramento Daily Union*, October 26, 1868.

69. *Frank Leslie's Illustrated Journal* printed the text from the telegrams, saying, "We publish the dispatches *as received*, the details, though necessarily meagre, showing that the damage has been considerable," as if the raw text of the telegrams had some kind of inherent truth buried in their unedited form. *Frank Leslie's Illustrated Journal*, November 7, 1868. Versions of the same dispatches from San Francisco appeared in newspapers such as the *Chicago Tribune* and *New York Times* on the day after the earthquake, October 22, 1868.

70. See, for example, the *Daily Milwaukee News*, October 22, 1868; *Elyria Independent Democrat*, October 28, 1868; *Chicago Tribune*, October 22, 1868.

71. "California," *Chicago Tribune*, October 22, 1868.

72. *Atlanta Constitution*, October 23, 1868; *Coshocton Democrat*, November 3, 1868; *Indiana Democrat*, November 5, 1868; *Anglo American Times*, October 24, 1868; *Edinburgh Evening Courant*, November 3, 1868.

73. "California," *New York Times*, October 22, 1868.

74. "Local Intelligence: Special Meeting of the Board of Supervisors," *Daily Alta California*, October 22, 1868.

75. "California," *New York Times*, October 22, 1868; "California," *Chicago Tribune*, October 22, 1868.

76. "Local Intelligence: The Great Earthquake of 1868." This also appeared in the *Daily Morning Call* and other newspapers on the same day.

77. See, for example, "Nearly Four Millions," *Daily Alta California*, October 22, 1868. The *Daily Morning Chronicle* claimed that it was "impossible to estimate the damage" but it "cannot fall short of $3,000,000." "Our Great Calamity."

The *Sacramento Daily Union* reported that "though the damages to property were at first roughly estimated at 'several millions,' the result was not so fearful nor so destructive of life as was apprehended." In in the same paragraph, however, it also concluded, "The list of buildings thrown down or damaged foots up a large number—large enough to justify a fear that the first reports of losses may not have been exaggerated." "Earthquake," *Sacramento Daily Union*, October 22, 1868.

78. The estimates of $300,000 were the only ones that were reported in *Coshocton Age*, October 30, 1868, and *St. Joseph Herald* (St. Joseph, MI), October 31, 1868.

79. *Fort Wayne Daily Gazette*, October 24, 1868. Other newspapers that contained ranges of estimates include the *Janesville Gazette*, October 24, 1868; *Daily Magnet* (Decatur, IL), October 26, 1868; *Dubuque Daily Herald*, October 25, 1868; "California," *Chicago Tribune*, October 25, 1868.

80. Richard T. Blackburn, "On the San Francisco Earthquake," *London Times*, November 21, 1868.

81. "San Francisco: The Great Earthquake; Full Particulars of the Event and its Results; From Our Special Correspondent," *Chicago Tribune*, November 5, 1868.

82. "California," *New York Times*, November 9, 1868.

83. *Journal of the Telegraph* 2, no. 1 (December 1, 1868): 7.

84. On the defense of the figures, see Wollenberg, "Life on the Seismic Frontier." On the inaccuracy of the damage estimates, see "Extravagant Estimates," *Oakland Daily Transcript*, October 26, 1868.

85. "Truth the Best Policy."

86. "Losses by the Earthquake," *Sacramento Daily Union*, November 2, 1868.

87. The *Union* further pointed out, "We prudently withheld their information till it had been confirmed" that the "ridiculously low estimate made out by the local papers and the parties assuming to inform the Eastern public" were incorrect. Ibid.

88. Ibid.

89. *The Great Earthquake in San Francisco: Estimated Damages in Detail* (San Francisco: White and Bauer, 1868). The report is not independent of the newspapers, particularly the *Bulletin*. The pamphlet authors used the same estimation techniques as the *Bulletin*, quoted the *Bulletin* in their description of the earthquake, and finally, in the

following sentence, observed that "we have given our readers a full and particular account in the *Bulletin*." Ibid., 17.

90. Ibid., 3–4. The *Sacramento Daily Union* had been arguing with estimates of $350,000 to $400,000 made by *Bulletin* journalists in the newspaper. "Losses by the Earthquake."

91. *Great Earthquake in San Francisco*, 16.

92. Wollenberg, "Life on the Seismic Frontier," 502. For example, a modern-day US Geological Survey paper uses the estimate of $350,000 worth of damages. Stover and Coffman, *Seismicity of the United States*, 439.

93. "For Eastern Friends," *Daily Alta California*, October 27, 1868.

94. Henkin, *The Postal Age*.

95. "Account of the Great Earthquake!," *Daily Alta California*, October 28, 1868.

96. "For Eastern Friends." The three telegraphic dispatches in *Frank Leslie's Illustrated Newspaper* appeared alongside a number of drawings of buildings in San Francisco before there was any damage. The pictures were views of the city hall and "plaza looking to the city hall." In the November 28, 1868, edition, *Harper's Magazine* published sketches of the damaged San Leandro courthouse.

97. "Advertisements," *Daily Alta California*, October 28, 1868.

D. Appleton and Company, "Earthquake in San Francisco, Oct. 21st 1868," reproduced in *The Henry H. Clifford Collection: Part Three, California Pictorial Letter Sheets*, prepared for auctions on October 26, 1994 (Dorothy Sloan, Rare Books, Austin, TX, 1994), item 54, plate 23; George H. Baker, "San Leandro Courthouse, Alameda Co. as Left by the Earthquake of Oct. 21, 1868," in Joseph Armstrong Baird Jr., *California's Pictorial Letter Sheets: 1849–1869* (San Francisco: David Magee, 1967), item 253.

98. Appleton's also advertised in the *San Francisco Daily Morning Chronicle*, October 24–27, 1868.

99. Quote from Vaughan's advertisement in the *Daily Alta California*, October 25, 1868. Other mentions of Vaughan's photography in "Local Intelligence," *Daily Alta California*, October 22, 1868.

Some of Eadweard J. Muybridge and Vaughan's photos are of the most heavily damaged buildings, and are currently held in archives and museums. Many are visible through the Online Archive of California (e.g., Eadweard J. Muybridge, "Effect of Earthquake in San Francisco, 21 Oct. 1868," San Francisco Bay Area Earthquake of 1868 Collection, Bancroft Library, University of California at Berkeley, http://ark .cdlib.org/ark:/13030/tf0199p0n9; Hector W. Vaughan, "Ruins of Unidentified Building: Following Earthquake of 1868, San Francisco," San Francisco Bay Area Earthquake of 1868 Collection, Bancroft Library, University of California at Berkeley, http://ark.cdlib.org/ark:/13030/tf4199p24r). There were also stereoptic images taken of the earthquake. See, for example, Isaiah W. Taber, "Effects of the Earthquake, Oct. 21, 1868, Railroad House, Clay St.," San Francisco Bay Area Earthquake of 1868 Collection, Bancroft Library, University of California at Berkeley, http://ark .cdlib.org/ark:/13030/tf9779p5rc.

100. Illustrated Earthquake Edition, *San Francisco Daily Morning Chronicle*, October 28, 1868. This special edition was also advertised in the *Daily Alta California* and *San Francisco Daily Morning Chronicle* on October 27, 1868.

101. "Our Great Calamity." Two days later, the *Daily Alta California* noted, "The photographers were at work before the buildings had fairly ceased shaking," and "the photographs are neatly executed, and give a most excellent idea of the present appearance of the buildings referred to." "Local Intelligence: After the Earthquake."

102. "Local Intelligence: After the Earthquake." The images were by photographers such as Muybridge, who published under the pseudonym "Helios." "Photograms of the Earthquake Scenes," *Daily Alta California*, October 28, 1868; "Earthquake Views," *Sacramento Daily Union*, October 28, 1868; Eadweard J. Muybridge, "Verso," San Francisco Bay Area Earthquake of 1868 Collection, Bancroft Library, University of California at Berkeley, http://ark.cdlib.org/ark:/13030/tf796nb86w. Rebecca Solnit also describes Muybridge's early work in *River of Shadows: Eadweard Muybridge and the Technological Wild West* (New York: Viking, 2003).

103. "San Leandro and Hayward's," *Daily Alta California*, October 27, 1868, quoting the *Oakland Daily News*, October 26, 1868.

104. In the years when California was becoming a state, in the midst of the gold rush, 1849–1851, six major fires swept through San Francisco.

105. "Extra!," *Oakland Daily News*, October 21, 1868.

106. *Oakland Daily Transcript*, October 22, 1868.

107. William Martis, "Caution regarding Chimneys," *San Francisco Evening Bulletin*, October 21, 1868. The *Daily Alta California* reported of a fire that "was caused by an improperly constructed chimney, which, being crushed by the earthquake, allowed the fire to communicate with the woodwork." The newspaper noted that the family was of "Spanish or Italian descent," which begs the question of whether they spoke English or if the warnings printed in the papers reached non-English-speaking communities in San Francisco. "Fire Yesterday Afternoon," *Daily Alta California*, October 23, 1868.

108. J. D. B. Stillman, "Concerning the Late Earthquake," *Overland Monthly and Out West Magazine* 1, no. 5 (November 1868): 479. This sentiment was echoed in many other newspapers: "Some Facts about Earthquakes," *Daily Alta California*, October 24, 1868; "Unsafe Buildings," *San Francisco Daily Morning Chronicle*, October 22, 1868; Editorial, *San Francisco Morning Call*, October 22, 1868.

109. "Local Intelligence: The Great Earthquake of 1868" (emphasis in original).

110. "Local Intelligence," *San Francisco Daily Morning Chronicle*, October 25, 1868.

111. "Danger from Insecure Buildings," *San Francisco Daily Morning Chronicle*, October 25, 1868.

112. The *Alta* wanted a superintendent of buildings, while the *Overland Monthly* advocated for a "commission with power to command ... all buildings and walls or chimneys endangering life, private as well as public." "Repairing Damages," *Daily Alta California*, October 24, 1868; Stillman, "Concerning the Late Earthquake," 479.

113. The first statewide antiseismic laws were passed in 1933 after an earthquake in Long Beach destroyed several (thankfully empty) school buildings.

114. "Local Intelligence," *San Francisco Daily Morning Chronicle*, October 22, 1868.

115. "Board of Education," *Daily Alta California*, October 25, 1868.

116. "Board of Education," *San Francisco Daily Morning Chronicle*, October 25, 1868; "Board of Education," *Daily Alta California*, October 25, 1868.

117. J. Denman, "The Public Schools," *Daily Alta California*, October 22, 1868.

118. "The Custom House, a brick building on pileground was badly shattered by the earthquake of October, 1865, is considered unsafe [after the 1868 earthquake]." "California," *Chicago Tribune*, October 22, 1868.

119. "United States Engineers Condemn the Custom-House Building," *San Francisco Daily Morning Chronicle*, October 24, 1868; "Removal of the U.S. Custom House," *Daily Alta California*, October 24, 1868; "The Custom House Building," *Daily Alta California*, October 25, 1868.

120. "Custom House Building."

121. Colonel Mendell, "Editor Bulletin," *San Francisco Evening Bulletin*, October 26, 1868.

122. "The Custom House," *San Francisco Evening Bulletin*, October 26, 1868.

123. On comparing San Francisco to other locales experiencing earthquakes and comparing earthquakes to other disasters, see "Some Facts about Earthquakes." On nihilism in San Francisco, see "An Unsafe Planet," *Daily Morning Chronicle*, October 23, 1868. On the assurance that this would be the worst California earthquake, see *Oakland Daily News*, October 22, 1868; "The Great Earthquake: Worldwide Disturbance," *San Francisco Evening Bulletin*, October 21, 1868.

124. "A gentleman who was in Callao during the recent earthquake in that city, says that the shock yesterday in San Francisco was equal in violence but much shorter in duration." W. H. H., "My First Experience of an Earthquake," *Sacramento Daily Union*, October 23, 1868.

"They seemed to feel that from flood and fire there was escape, but from earthquakes—especially like that of which they were doubtless reminded in South America—none." "City Intelligence at San Francisco." The report was from "a gentleman who came up from San Francisco last evening." One letter writer noted that people could not help but remember the stories of the devastation of the earthquake in South America: "This introduction is immediately induced from the burning memory of what a sister continent has just experienced." H. P., "The Great Earthquake."

125. "Some Facts about Earthquakes."

126. Stillman, "Concerning the Late Earthquake," 474 (emphasis in original).

127. C. W. C., "San Francisco, Nov. 9," *Chicago Tribune*, November 25, 1868; Geschwind, *California Earthquakes*, 16.

128. Ibid., 12; Peggy O'Donnell, "Earthquakes: An Introduction," *Synergy* 1, no. 15 (March 1969): 4. In the California Academy of Science's *Proceedings*, Trask published at least ten articles in his career. The first, in 1856, was a catalog of earthquakes in California from 1812 to 1856. The most recent were John Boardman Trask, "Earthquakes in California from 1800 to 1864," "Earthquakes in California during 1864," and "Earthquakes in California during 1865," in *Proceedings of the California Academy of Natural Sciences, Volume 3: 1863–1867* (San Francisco: Bacon and Company, 1868): 130, 190, 239.

129. California Academy of Science, *Stated Meeting Minutes, Jan., 1868–Jan., 1872*, November 2, 1868, 74, California Academy of Science Archives, San Francisco; Theodore Henry Hittell, *Theodore Henry Hittell's The California Academy of Sciences: A Narrative History, 1853–1906*, ed. Alan E. Leviton and Michele L. Aldrich (San Francisco: California Academy of Sciences, 1997).

130. California Academy of Science, *Stated Meeting Minutes*.

131. Hittell, *California Academy of Sciences*.

132. California Academy of Science, *Stated Meeting Minutes*.

133. Washington Bartlett, "Report of the Secretary to the 20th Annual Meeting of the Chamber of Commerce of San Francisco," in *Annual Reports to the Chamber of Commerce of San Francisco, Submitted to a Meeting of the Chamber, Holden May 10th, 1870: To Which Are Appended[:] Report from Sub-Committee of the Committee on Earthquakes, Report of the Committee on Dockage and Wharfage, Report of the Committee on United States District Courts, Report of the Committee on Laws Relating to Pilots and Pilotage, and Also, Memorials and Resolutions Adopted by the Chamber, during the Year Ending May 10, 1870* (San Francisco: Alta California Printing House, 1870), 9.

134. This report has been described as "missing" in modern scholarship and has generated great intrigue. There has been much written about the existence or

nonexistence of a report giving details about the 1868 earthquake, and whether a report was suppressed because of a letter from George Davidson, a well-reputed surveyor, who claimed that the committee head, George Gordon, suppressed the report because it would reflect poorly on the business prospects for the city—a sentiment reflected in the famous Lawson report that analyzed the 1906 earthquake. William H. Prescott, "Circumstances Surrounding the Preparation and Suppression of a Report on the 1868 California Earthquake," *Bulletin of the Seismological Society of America* 72, no. 6A (December 1, 1982): 2389–2393; Michele L. Aldrich, Bruce A. Bolt, Alan E. Leviton, and Peter U. Rodda, "The 'Report' of the 1868 Haywards Earthquake," *Bulletin of the Seismological Society of America* 76, no. 1 (1986): 71–76; Lawson et al., *California Earthquake of April 18, 1906*.

In fact, researchers in several fields have concluded that the report was underfunded, California lacked the people with adequate scientific knowledge to produce such a report, and the organization in charge of reporting collapsed due to the death of the committee head and unexplained loss of the records that he had been collecting (but not sharing widely). California was not a hotbed of seismology knowledge in 1868—unlike in Japan and Europe, where scientists were developing serious research programs about earthquakes. Geschwind has said that there was not enough scientific expertise in California to produce a report:

> Even though San Francisco was a booming metropolis ... scientific institutions there were still too weak to support a sustained investigation: the Academy of Sciences comprised mostly amateur members; the newly chartered University of California had not yet opened its doors; and the state geologist, disappointed at the lack of financial support for his survey, had just returned to a teaching position at Harvard. Under these circumstances, the study of the 1868 earthquake descended into bickering among supporters of various speculative theories.

Geschwind, *California Earthquakes*, 12–13.

Tobriner argues that the disparaging remarks about the qualifications and lack of scientific expertise of the committee, made by Thomas Rowlandson, the secretary of the Earthquake Committee, might have contributed to the delayed report.

Gordon, as the head of the Joint Earthquake Committee, wrote vigorously about the need to synthesize the experiences of the public in a report, inviting contributions from the earthquake public. But Gordon died in May 1869, not having produced the report. Bartlett, "Report of the Secretary"; Wollenberg, "Life on the Seismic Frontier," 507.

Both Wollenberg and Aldrich and colleagues cite Blake, a member of the subcommittee charged with scientific inquiry; Blake himself noted in the *Annual Reports to the San Francisco Chamber of Commerce* from 1870 that there was insufficient funding for the project. Wollenberg, "Life on the Seismic Frontier"; Aldrich et al., "The 'Report' of the 1868 Haywards Earthquake."

135. From its inception, the committee's goals were to have guidance that was practical for the public to use: "Resolved. That our researches and experiments shall be subsidiary to this practical result, and that the promotion of abstract science be considered as secondary and incidental thereto." There were five subcommittees

formed in late November 1868, each concerned with different aspects of earthquakes and structures: "bricks, stones, and timbers"; "limes, cements, and other bonds and braces"; "structural designs"; "scientific inquiry and collection of facts"; and "Legal—The Law Governing Building." "Local Intelligence," *Daily Alta California*, November 25, 1868.

136. "The Secretary of that Committee, Professor Rowlandson had published a pamphlet on that subject under the auspices of several members of the Committee. This work of Mr. Rowlandson, though not the official report, contained a large amount of valuable information on the subject of earthquakes." "Annual Chamber of Commerce Meeting," *San Francisco Daily Morning Chronicle*, May 12, 1869.

137. "Strength of Buildings," *Daily Alta California*, May 15, 1869. Details about California were scant in the Rowlandson pamphlet because apparently Rowlandson was told not to comment on the California earthquake following his dismissal from the committee. On the title page of his book, he refers to himself as "Fellow of the Geological Society, London, and *Late* Secretary of the Joint Committee on Earthquake Topics" (emphasis added). Thomas Rowlandson, *A Treatise on Earthquake Dangers, Causes, and Palliatives* (San Francisco: Dewey and Co., 1868).

138. Ibid., 4. I derived the date and background on Rowlandson from Aldrich et al., "The 'Report' of the 1868 Haywards Earthquake."

139. Deborah R. Coen, *The Earthquake Observers* (Chicago: University of Chicago Press, 2013), 187–201.

140. James Blake, "Report on the Subcommittee on Earthquake Topics," in *Annual Reports to the Chamber of Commerce of San Francisco, Submitted to a Meeting of the Chamber, Holden May 10th, 1870 : To Which Are Appended[:] Report from Sub-Committee of the Committee on Earthquakes, Report of the Committee on Dockage and Wharfage, Report of the Committee on United States District Courts, Report of the Committee on Laws Relating to Pilots and Pilotage, and Also, Memorials and Resolutions Adopted by the Chamber, during the Year Ending May 10, 1870* (San Francisco: Alta California Printing House, 1870), 14.

141. "Why Not Report," *San Francisco Daily Morning Chronicle*, August 28, 1869.

142. "Chamber of Commerce," *Daily Alta California*, May 11, 1870; "Chamber of Commerce Meeting," *San Francisco Daily Morning Chronicle*, May 11, 1870; *Annual Reports to the Chamber of Commerce of San Francisco.*

143. Bartlett, "Report of the Secretary," 10.

144. Observer, "The Earthquake Investigation," *Daily Alta California*, December 7, 1868.

145. "Scholars and lay people alike had a new awareness of seismic danger and of earthquake-resistant retrofit and design." Tobriner, *Bracing for Disaster*, 58.

146. The earthquake "repeats with terrible emphasis the warning the people of this city had three years ago … to build strongly, on strong foundations; to iron brace ever the stoutest walls, and to avoid heavy projecting cornices and loosely attached ornamentation." The article goes on to say that even though the earthquake was part of larger phenomena, the "local lessons it teaches" should be attended to. *San Francisco Evening Bulletin*, October 21, 1868. See also "The Results and Lessons of the Earthquake," *Daily Alta California*, October 22, 1868; W., "Local Intelligence," *Daily Alta California*, October 26, 1868; letter addressed "to Editors Alta," *Daily Alta California*, October 24, 1868.

147. "Local Intelligence: The Great Earthquake of 1868."

148. "The Real Estate Market," *San Francisco Daily Morning Chronicle*, October 25, 1868.

149. "Prosy 'Lessons,'" *San Francisco Daily Morning Chronicle*, October 25, 1868.

150. *Scientific American* 19, no. 22 (November 25, 1868): 342.

151. "I presume you have read an account in the papers of the earthquake." Nettie Denman to cousin, December 10, 1868, box 5, file 10, Small Manuscript Collection, San Francisco Public Library. See also J. McDowell to Henry A. Collin, October 21, 1868, BANC MSS 2005/196c—BANC FILM 3233, Henry A. Collin Correspondence, 1856–1875 Collection, Bancroft Library, University of California at Berkeley.

152. John, *Network Nation*.

Chapter 3

1. Dorothy H. Fowler, *A Most Dreadful Earthquake: A First-Hand Account of the 1906 San Francisco Earthquake and Fire, with Glimpses into the Lives of the Phillips-Jones Letter Writers* (Oakland: California Genealogical Society, 2006), 29–30, letter no. 10, April 27, 1906.

2. Sarah's father and George were cousins; Sarah and George were later married. Sarah survived the earthquake unharmed. Initially Sarah's flat survived the earthquake. It was later destroyed in the fire, however, and she had to move to her friend's house on Stanyan. This account of Sarah and George is based on a collection of their letters edited and published by the California Genealogical Society. Fowler, *Most Dreadful Earthquake*.

3. The population grew from about 150,000 (in 1870) to 350,000 (in 1900), as had the population of the San Francisco Bay Area, reaching approximately 660,000 in 1900. Metropolitan Transportation Commission and the Association of Bay Area Governments, Bay Area Census, accessed July 30, 2017, http://www.bayareacensus.ca.gov.

The circulation of the largest newspapers skyrocketed during this time. The *Bulletin's* circulation grew from 11,000 to 58,045 from 1870 to 1906, and the *Chronicle* went from 25,200 to 80,626. Emerson L. Daggett, ed., *Trends in Size, Circulation, News, and Advertising in San Francisco Journalism* (San Francisco: Works Progress Administration of Northern California, 1940), 4:65.

Additionally, the telegraph infrastructure had grown and the telephone had been introduced. While sending telegrams was a regular business practice, it was still expensive, although more widely used than in 1868.

4. Fowler, *Most Dreadful Earthquake*, 11, letter no. 4, April 18, 1906.

5. Ibid., 14, letter no. 5, April 18, 1906.

6. Ibid., 38, letter no. 12, April 28, 1906. Sarah's companion also reassured George that Sarah did not want him wondering about her well-being based on newspaper reports: "She was so anxious to have the telegram reach you before you would see an account of it in the paper." Ibid., 34, letter no. 11, April 27, 1906.

7. Ibid., 36, letter no. 12, April 28, 1906.

8. Ibid., 27, letter no. 9, April 25, 1906; ibid., 37, letter no. 12, April 28, 1906.

9. Ibid., 24, letter no. 8, April 23, 1906. Sarah did not have to "take advantage of the privilege" of sending mail without postage because she had saved letters and stamps for mail when she abandoned her flat.

10. Ibid., 15–20, letters nos. 6 and 7, April 23, 1906.

11. Ibid., 17–18, letter no. 6, April 23, 1906.

12. Ibid., 26, letter no. 9, April 25, 1906. This letter gives hints as to how Sarah and her acquaintances attempted to locate Lizzie: "Mr. Rush has searched for her and we have advertised and written letters but we have heard nothing. Tomorrow, I will begin a systematic search, going to each one of the registry bureaus." Lizzie was eventually found; she was in a hospital after stepping on a nail.

13. "Message in a Loaf of Bread," *San Francisco Chronicle*, April 30, 1906; Richard Schwartz, *Earthquake Exodus, 1906: Berkeley Responds to the San Francisco Refugees* (Berkeley: RSB Books, 2005), 86; "Notes Hidden in Loaves of Bread," *Berkeley Daily Gazette*, April 26, 1906.

14. One example of a posted notice read, "To be found at about 250 yds. West of Children's Playground, G. G. Park, Bernice Blacklock, Freddie Blacklock, Leona Blacklock, Mrs. Irene R. Smith, J. J. Smith, Joe Schaeffer, Arthur Moore." "The Great Fire of 1906, L: Adventures in Finding Lost Friends and Relatives—A Fence-Post Directory," *Argonaut*, April 2, 1927, http://ark.cdlib.org/ark:/13030/hb7n39p2jt/?order=6. For other examples, see Schwartz, *Earthquake Exodus*, 84.

15. "S.F. after 1906 Fire [and] Earthquake. [Refugees at bulletin board in camp, unidentified location]," 1906, image, California Historical Society, San Francisco, http://ark.cdlib.org/ark:/13030/hb6n39p1w6.

16. Businesses proclaimed their services however they could. A horse with a carriage attached was draped with a sign that said "safes opened," parked in front of a sign advertising safe removal and opening as well as building wrecking services. A photograph of Fillmore Street, the main commercial street after the earthquake, shows a number of sign-making shops prominently advertised. RG111-SC, Records of the Office of the Chief Signal Officer—Prints—Military History 1860–1938, box 732, photos numbered SC-95230, National Archives, College Park, MD; Records of the Bureau of Public Roads—Slides—Photographs of the Aftermaths of the San Francisco Earthquake April 1906, box 1, slide number 30-HH-4, National Archives, College Park, MD.

17. This seismic denial argument is advanced in several works: Gladys Hansen and Emmet Condon, *Denial of Disaster: The Untold Story and Photographs of the San Francisco Earthquake of 1906* (San Francisco: Cameron and Company, 1989), 109–110; Carl-Henry Geschwind, *California Earthquakes*: Science, Risk, and the Politics of Hazard Mitigation (Baltimore: Johns Hopkins University Press, 2001), 20–23; Ted Steinberg, *Acts of God*: The Unnatural History of Natural Disaster in America, 2nd ed. (New York: Oxford University Press, 2006), 30.

From the perspective of the twenty-first-century researcher, newspapers provide an invaluable record, used in many popular and academic accounts. As my chapter describes, however, some newspapers were involved with other Bay Area elites in rebranding the "San Francisco Earthquake and Fire" as the "San Francisco Fire" to protect their (and others') ability to collect on their fire insurance. Furthermore, newspapers during this period had a reputation for sensationalism and muckraking. The newspapers were not always reliable as sources for anything but documenting what newspapers said. A distrust of the newspaper might lead one to focus on letters and personal accounts describing the disaster, but these also must be read with an eye to ideological and political campaigns, because ordinary citizens were invited to participate in the rebranding campaign. Hansen and Condon, *Denial of Disaster*, 107–134. Several historians, including Geschwind and Steinberg, argue that there was a concerted campaign to minimize the amount of damage done by the earthquake and deny that earthquakes could harm Californians in the future—similar to what happened with nineteenth-century earthquakes, like the one in 1868. As Geschwind notes, "Californians reacted" to the earthquake and fire by relying "on patterns of behavior established in response to previous earthquakes." Geschwind, *California Earthquakes*, 21; Steinberg, *Acts of God*, 26–36; Ted Steinberg, "Smoke and Mirrors: The San Francisco Earthquake and Seismic Denial," in *American Disasters*, ed. Steven Biel (New York: NYU Press, 2001), 103–128.

18. Many secondary sources attempted to summarize the earthquake or mine it for profit. There were eighty-two popular books written about the 1906 earthquake for

popular consumption immediately after the earthquake. These books were generally more sensationalized, and some authors never actually visited the earthquake site. Twenty-first-century historians used the popular works as examples of pop culture renditions of the earthquake. A number of other popular accounts were published since the earthquake. Philip Fradkin, *The Great Earthquake and Firestorms of 1906: How San Francisco Nearly Destroyed Itself* (Berkeley: University of California Press, 2005), 263–288; Kevin Rozario, *The Culture of Calamity*: Disaster and the Making of Modern America (Chicago: University of Chicago Press, 2007).

19. Andrea Rees Davies, *Saving San Francisco: Relief and Recovery after the 1906 Disaster* (Philadelphia: Temple University Press, 2011); Marian Moser Jones, *The American Red Cross from Clara Barton to the New Deal* (Baltimore: Johns Hopkins University Press, 2013). After the earthquake, many organizations published reports about their activities during the disaster, including the major constituents involved with relief activities: the military, Red Cross, and civilian elites in charge of the Finance Committee; many other fraternal or charitable organizations issued reports on the activities of their members. Some of these reports have been treated with suspicion by researchers in the twenty-first century and viewed as opportunities for institutional aggrandizement. This bibliographic insight tells us not just about the number of people involved but also, as many scholars argue, the Progressive mood of the era—a focus on record-making and using data in decision-making. One achievement of these Progressive ideals is the 1913 Russell Sage Foundation report called *The San Francisco Relief Survey: The Organization and Methods of Relief Used after the Earthquake and Fire of April 18, 1906*. It is a comprehensive analysis of the relief project from the perspective of Progressive reformers. Another notable document created in the wake of the earthquake is *The California Earthquake of April 18, 1906: Report of the State Earthquake Investigation Commission in Two Volumes and Atlas* (what is now called the Lawson Report). It was assembled by a coalition of California seismologists and published by the Carnegie Institution. Today, it is remembered as the first in-depth US study of an earthquake and an important landmark for California seismology. The volume of material produced by various institutions about the earthquake is illustrative of the massive number of people involved in the disaster response.

20. "Crowds Begging Temblor News," *Los Angeles Herald*, April 19, 1906; "Chicago People in Much Anxiety," *Chicago Daily Tribune*, April 19, 1906; "Inquiries Swamp Telegraph Lines," *Chicago Daily Tribune*, April 20, 1906.

21. A University of California at Berkeley professor gathered these recollections in 1919. Many of the students were quite young at the time, so their stories are of dubious quality. There are, however, many of them, and they paint a broad picture of information practices that is hard to get elsewhere, especially as experienced within families. Professor George Stratton's assignment reads as follows: "Earthquake of 1906: Where were you at the time? Write as detailed and precise an account as you can of your own recollections of the earthquake and fire of 1906. Tell if you remember them, the events of the day and evening before the eathquare [*sic*], the doings

and thought of yourself during and after the earthquake. the [sic] doings of your family and neighbors,—all these for the days and weeks following until the earthquake no longer figures in the events. Tell nothing *that you do not personally remember*. If you remember little tell that little carefully." George Malcom Stratton, Recollections of the San Francisco 1906 Earthquake by Stratton Students, 1919, Bancroft Library, University of California at Berkeley, BANC MSS C-B 1032, carton 6.

22. W. Kennedy, living in Rocky Ford, Colorado, in ibid., folder 1.

23. M. L. Gelber, living in New York City, in ibid., folder 1.

24. Corinne Connell, living in San Diego, in ibid., folder 1.

25. Additionally, the information infrastructure overlapped with the financial infrastructure; money could be "wired," and thus those who had lost everything might have been particularly eager to pick up telegrams, given that it was "better to receive a message back from a friend that has some money to send. ... Thousands of messages have flashed across the wires since that memorable morning ... the accumulated messages have crept well up into hundreds at times during the past few days." Alice G. Eccles, "Telegraph Wires Are Laden with Messages," *Oakland Tribune*, April 22, 1906.

26. Record Group 59, General Records of the Department of State, Miscellaneous Correspondence, 1784–1906, Special Series of Domestic and Miscellaneous Messages of Condolence, Official Messages on the San Francisco Earthquake, April 19–25, 1906, box 1; NARS A-1; Entry 182, National Archives, Washington, DC.

27. Apparently a dramatic last telegraph was sent out: "The city practically ruined by fire. It's within half a block of us. ... The Call Building is burned out entirely. The Examiner Building just fell in a heap. ... They are blowing standing buildings that are in the path of flames with dynamite. No water. It's awful. There is no communication anywhere and entire phone system is busted. I want to get out of here or be blown up." Chief operator, Postal Telegraph Office, San Francisco, 2:20 p.m., quoted in William Bronson, *The Earth Shook, the Sky Burned: A Photographic Record of the 1906 San Francisco Earthquake and Fire* (San Francisco: Chronicle Books, 2006), 56.

28. Eccles, "Telegraph Wires Are Laden with Messages."

29. A. J. Esken and J. W. Whitley, "Personal Experience of the Telegraphers at San Francisco at the Time of the Earthquake and Fire," *Telegraph Age* 24, no. 12 (June 16, 1906): 283–285.

30. Paul Cowles, "The Associated Press Men's Work at San Francisco—How the Earthquake News Was Gathered and Sent Out," *Telegraph Age* 24, no. 12 (June 1, 1906): 252–253.

31. "Telegraph Office Perched on Pole—How the Western Union Built a New Plant in Four Days," *San Francisco Chronicle*, April 30, 1906.

32. H. J. Jeffs, "The Western Union in the San Francisco Disaster," *Telegraph Age* 24, no. 12 (June 1, 1906): 254–255.

33. Ibid.; "Telegraph Office Perched on Pole."

34. Jeffs, "Western Union in the San Francisco Disaster"; "Telegraph Office Perched on Pole."

35. Jeffs, "Western Union in the San Francisco Disaster."

36. "At 3 o'clock it had got to the Palace Hotel on the Mission-Street side, and by 3:30 it was well on fire. About this time I went into the Western Union Telegraph office, and while writing a telegram to Nellie and Robert, who were on their way to New York, the announcement was made that no more telegrams would be received." James B. Stetson, San Francisco during the Eventful Days of April, 1906: Personal Recollections, 1906, 10, Bancroft Library, University of California at Berkeley.

37. All the newspaper companies went without reliable telegraph lines into San Francisco for a week, having reporters carry stories across the San Francisco Bay on boats to Oakland to be printed or sent to the rest of the country. "Newspapers Show Great Resources: Under Incredible Difficulties All but One of the San Francisco Dailies Survive," *San Francisco Bulletin*, April 29, 1906; "S.F. Telegraph Office's First Stand under a Roof in Shack on Oakland Water Front" (caption on print: "2nd office in shack. Oakland"), 1906, image, California State Library. http://ark.cdlib.org/ark:/13030/hb5d5nb3tt.

The *San Francisco Chronicle* summarized its technical needs required to inform people of the outside world: "What the press can do toward satisfying this desire will depend on the mechanical possibility of the telegraph and the available printing presses." Russell Quinn, *The San Francisco Press and the 1906 Fire*, ed. Emerson L. Daggett (San Francisco: Works Progress Administration of Northern California, 1940), 5:30, quoting the *San Francisco Chronicle*, April 26, 1906.

38. "Telegraph Office Perched on Pole."

39. "Newspapers Show Great Resources."

40. "Telegraph Office Perched on Pole."

41. Paul Cowles, superintendent of the western division of the Associated Press, said, "Monday night [after the earthquake] we secured a wire from the telephone company, which we turned into a Morse circuit. We used this wire for two nights, which were nights of joy, but on the third night the telephone company went out of business, their cable having been blown up, and we were forced back to the old system of delivery by messenger to Oakland." Cowles, "Associated Press Men's Work at San Francisco," 253.

See also "Long-Distance Phones Work," *San Francisco Bulletin*, April 23, 1906. "The wireless" was also used to deliver messages. See "The Great Fire of 1906, LII:

How the Wireless Helped—Unique Achievement of the "Daily News," *Argonaut*, April 16, 1927.

42. Photographs show offices located in makeshift shacks in burned areas of the city. See, for example, photograph RG111-SC, Records of the Office of the Chief Signal Officer—Prints-Military History, 1860–1938, box 732, photograph number SC-95177, National Archives, College Park, MD; "Telegraph Office in Election Booth on Van Ness Avenue," 1906, image, California State Library, http://ark.cdlib.org/ark:/13030/hb5d5nb3tt; "Makeshift Office of Postal Telegraph Co. Commercial Cables, in Front of Their Ruined Building, Market St.," image, California Historical Society, San Francisco, http://ark.cdlib.org/ark:/13030/hb796nb67f. A description is also found in Jeffs, "Western Union in the San Francisco Disaster."

43. Cameron King Jr., deputy registrar of voters at City Hall, to Anna Strunsky Walling. They met through the Socialist Labor Party. Mary McD. Gordon and Cameron King Jr., "Earthquake and Fire in San Francisco," *Huntington Library Quarterly* 48, no. 1 (1985): 69–79; "Overcharging for Telegrams," *San Francisco Chronicle*, April 22, 1906. Allegations of overcharging for telegrams appear to be justified.

44. "Crowds Trying to Telegraph Relatives from Oakland Offices," 1906, image, California State Library, http://ark.cdlib.org/ark:/13030/hb5z09p13z.

45. "Telegraph Office Perched on Pole."

46. "He must know at a glance the matter that should be put ahead and then he must see that it is routed with the least possible delay and in the most direct way." Ibid.

47. "Impossible to Send Messages," *Los Angeles Herald*, April 21, 1906.

48. "Carries Dismay to Sister City," *San Francisco Call*, April 22, 1906, from the *Los Angeles Times*.

49. Chicago apparently had been able to get a cable to the Ferry Building in San Francisco. In Chicago, priority was given to the news coming out of San Francisco rather than sending messages there. "Chicago People in Much Anxiety."

50. "Seek Tidings from Home," *Chicago Daily Tribune*, April 19, 1906.

51. "The Telegraph Companies Refute the Action of the San Francisco Grand Jury," *Telegraph Age* 24, no. 13 (July 1, 1906): 301–302.

52. "Messages Sent North upon the Owl," *San Francisco Bulletin*, April 21, 1906.

53. James A. Warren, telegram, April 20, 1906, MS 3461, California Historical Society, San Francisco, http://ark.cdlib.org/ark:/13030/hb2d5nb1gr.

Reports about the volume of telegraphs are present in many newspaper articles and in letters as well: "Telegrams could be sent from Oakland, but it was almost impossible to get over there, and then almost as impossible to get a telegram sent as thousands and thousands of telegrams were filed and as many received. It is said that 10,000 telegrams alone were received by mail from outside points, at Oakland, and they say that the Western Union at Chicago took 15,000 telegrams they could not get on the wire, so put a messenger on the train and sent him through with them. Mr. Shields received a telegram from Seattle on the 27th. That was filed there on the 18th. Mr. Brick, one of our travelers received a telegram here in the office on the 30th. In the morning from Newark, New Jersey, dated there the 18th. I tried to telegraph the first thing Wednesday morning but was informed the wires were down, and then immediately the fire got full sweep and the mails were out of commission." Charles E. Leithead, "Account of the 1906 San Francisco Earthquake and Fire," May 2, 1906, 11–12, California Historical Society, San Francisco, http://ark .cdlib.org/ark:/13030/hb6b69p1t6.

54. T. W. Goulding, Letter to J. P. Spanier: LS, London, April 24, 1906, BANC MSS C-Z 95, Bancroft Library, University of California at Berkeley, http://ark.cdlib.org/ ark:/13030/hb8x0nb6qx.

55. There are numerous examples of postmarked telegrams. See, for example, Arthur Dangerfield, Telegram, Postmarked April 23, 1906, BANC MSS 94/24 c 1:06, Bancroft Library, University of California at Berkeley, http://ark.cdlib.org/ark:/13030/ hb558007gz.

56. "Telegraph Companies Refute the Action of the San Francisco Grand Jury," 301.

57. "The committee [of the Grand Jury] believes that the company committed the grossest fraud in maintaining its sign purporting to be doing a telegraphic business, when in reality it was taking the people's money and sending messages by messenger or by mail." "Want to Indict Western Union," *San Francisco Call*, June 14, 1906.

58. A joint press conference from the heads of the Western Union Company and Postal Telegraph Company was the response, in which charges of fraud were countered. Quotes from the conference appeared in many venues friendly to the telegraph companies. See, for example, "Telegraph Officials Answer Fraud Charge," *New York Times*, June 16, 1906; "Telegraph Men Reply: Call Charges Wicked," *New York Daily Tribune*, June 16, 1906.

59. "Telegraph Companies Refute the Action of the San Francisco Grand Jury," 301.

60. All the accounts of the telegraph operators and managers note how long and hard employees worked to restore service. See, for example, Jeffs, "Western Union in the San Francisco Disaster"; "Telegraph Office Perched on Pole"; Cowles, "Associated Press Men's Work at San Francisco"; Esken and Whitley, "Personal Experience of the Telegraphers at San Francisco."

61. Gregory J. Downey, *Telegraph Messenger Boys: Labor, Technology, and Geography, 1850–1950* (New York: Routledge, 2002).

62. "Telegraph Companies Refute the Action of the San Francisco Grand Jury"; "Useless to Send Telegrams," *Chicago Daily Tribune*, April 21, 1906; T. W. Goulding, Letter to J. P. Spanier.

63. Downey, *Telegraph Messenger Boys*.

64. Ibid.; "Telegraph Companies Refute the Action of the San Francisco Grand Jury."

65. "Impossible to Send Messages"; "Useless to Send Telegrams."

Furthermore, a telegraph company executive noted that "martial law was established over San Francisco on the day of the fire, and no one was permitted to pass within the lines, so that it was impossible for a number of days to make any attempt to effect delivery of telegrams in San Francisco, even if it would have been possible to have located the persons to whom the telegrams were addressed." While many San Franciscans believed that they were under martial law, legally they were not. "Telegraph Companies Refute the Action of the San Francisco Grand Jury."

66. "Uncalled for Telegrams," *San Francisco Bulletin*, April 28, 1906; "Undelivered Telegrams," *San Francisco Bulletin*, April 29, 1906; "Telegraph Companies Refute the Action of the San Francisco Grand Jury."

67. Richard R. John, *Network Nation: Inventing American Telecommunications* (Cambridge, MA: Harvard University Press, 2010).

68. William F. Burke, "The Great Fire of 1906, XXXV: How the Post Office Rose Superior to the Disaster—Unstamped Cuffs as Mail Matter," *Argonaut*, December 18, 1926, http://ark.cdlib.org/ark:/13030/hb5c600874/?order=3&4.

69. Randy Stehle, "Auxiliary Markings: 'Burned Out' in the 1906 San Francisco Earthquake and Fire," *La Posta* 20, no. 6 (December 1989): 7–12; Randy Stehle, "Auxiliary Markings: 'Burned Out'; in the 1906 San Francisco Earthquake and Fire— Recent Discoveries and a Re-Examination of the Resumption of Normal Postal Service," *La Posta* 29, no. 3 (July 1998): 7–28. Both reprinted in Randy Stehle, *Postal History of the 1906 San Francisco Earthquake and Fire* (West Linn, OR: La Posta Publications, 2010): 7 and 12.

70. Before the earthquake, sometimes streetcars with "United States Mail" signs on them handled the mail. One of these signs was hung on an automobile on April 21, and driven around to where people were located to let them know that the post office would be picking up outgoing mail: "The effect was electrical. ... [People] shouted in a state bordering on hysteria. ... As we went on into the Presidio there was almost a riot. ... It was the same in the Park." Burke, "How the Post Office Rose Superior to the Disaster."

71. Ibid.

72. News spread quickly. "Letters Sent without Stamps," *San Francisco Chronicle*, April 21, 1906; "San Francisco Post Office Is Again at Work—Mails Being Distributed and Collected at Branch Stations for Refugees—Every Facility Afforded to the Public for Communication with Outside," *San Francisco Call*, April 22, 1906.

73. See, for example, the telegram pictured in figure 3.7: Hilda Blight letter to Mrs. A. E. Mitchell, April 22, 1906, MS 3463, California Historical Society, San Francisco, http://ark.cdlib.org/ark:/13030/hb429005w7.

74. "All Mail Held at Local Office," *San Francisco Chronicle*, April 22, 1906.

75. Burke, "How the Post Office Rose Superior to the Disaster."

76. Ibid.

77. Stehle, "Auxiliary Markings: 'Burned Out' in the 1906 San Francisco Earthquake and Fire," 8.

78. Randy Stehle, "The 1906 San Francisco Earthquake and Fire: Recent Discoveries (Part 3)," *La Posta* 33, no. 1 (March 2002): 47–50, reprinted in Stehle, *Postal History of the 1906 San Francisco Earthquake and Fire*, 34.

The image also appears as "Title not known," Gifford M. Mast, 1906, University of California at Riverside, Museum of Photography, http://content.cdlib.org/ark:/13030/kt6x0nc2gj/?order=2.

Pictures of a refugee tent at the Lobos Square refugee camp show a post office: "Sewing Class in Session at 'Forward Movement' Tent, Refugee Camp, Lobos Square" (photo from collection of Jesse B. Cook), 1906, FN-34241, California Historical Society, San Francisco, http://ark.cdlib.org/ark:/13030/hb5n39p151; "Group Portrait of Sewing Class at 'Forward Movement' Tent, Harbor View Refugee Camp," 1906, 1 FN-19909, California Historical Society, San Francisco, http://ark.cdlib.org/ark:/13030/hb109nb1c8.

79. Still, it is important not to overstate that notions of equality motivated the service of the post office. Many of the actions taken after the earthquake were ad hoc, with an eye to universal service, but these ad hoc actions were tempered by the same bureaucratic restrictions as before the earthquake.

80. Stehle, "Auxiliary Markings: 'Burned Out' in the 1906 San Francisco Earthquake and Fire," 8; Stehle, "Auxiliary Markings: 'Burned Out'; in the 1906 San Francisco Earthquake and Fire—Recent Discoveries and a Re-Examination of the Resumption of Normal Postal Service," 13, 15.

81. Camp Ingleside was for the elderly and infirm who were without people to care for them; built in former horse stables, it was apparently an undesirable place to end up. Davies, *Saving San Francisco*, 100–110.

82. Burke, "How the Post Office Rose Superior to the Disaster."

83. Ibid.

84. Jessica Lemieux, "Phoenix Rising: Effects of the 1906 Earthquake on California Print Culture" (Master's thesis, San Jose State University, 2006). Moreover, historians have maintained that these newspapers asserted their power by participating in a project to downplay and deny the earthquake's effects, emphasizing that the damage was done by fire (a type of disaster for which many had insurance). See Hansen and Condon, *Denial of Disaster*, 107–134; Geschwind, *California Earthquakes*, 21; Steinberg, *Acts of God*, 26–36; Steinberg, "Smoke and Mirrors." Rozario argues that narratives of renewal were encouraged because disaster meant a "clean slate" for improvement; there were even accounts that focused on the feelings of euphoria after the earthquake, but they were primarily by members of the middle class, who had various types of safety nets. Rozario, *Culture of Calamity*, 23.

85. The *Call-Chronicle-Examiner* collaboration of April 19 would not continue. The *Examiner* bribed the *Tribune* to have sole access to the presses, and it was another day before the other newspapers could publish. April 21, 1906, was really the first day that was "back to business" for the major San Francisco newspapers. The *Bulletin*, meanwhile, negotiated using the press of the *Oakland Herald* in the afternoons, while the *Call* also made use of the *Herald* presses. The *Chronicle* eventually went to the *Oakland Examiner*. Quinn, *San Francisco Press and the 1906 Fire*, 5:15–19.

86. Philip J. Ethington, *The Public City: The Political Construction of Urban Life in San Francisco, 1850–1900* (New York: Cambridge University Press, 1994).

87. "Hope Springs Eternal," *San Francisco Chronicle*, April 21, 1906.

88. "Many Register at Park Station," *San Francisco Call*, April 21, 1906.

89. "The following San Franciscans and others have registered at the 'Chronicle' business office, 1236 Broadway, Oakland, giving their Oakland or San Francisco address." "Registered at 'Chronicle:' Lists of San Franciscans and Others Who Give Their Addresses at the Present Time," *San Francisco Chronicle*, April 23, 1906.

90. "Parents and Children Separated," *Washington Times*, April 21, 1906.

91. "Registry Bureaus throughout City," *San Francisco Call*, April 21, 1906; "Hope Springs Eternal."

92. "Many Families Divided—Bureaus of Registry in Oakland Crowded by Anxious Persons," *New York Tribune*, April 23, 1906. "Illimitable and almost impossible" is also from an AP article appearing in other newspapers such as the *Los Angeles Times*, April 22, 1906.

93. "Oakland Houses 75,000 Refugees," *Los Angeles Herald*, April 23, 1906.

94. "Many Families Divided"; "Where to Register on Either Side," *San Francisco Chronicle*, April 22, 1906.

95. "List of Homeless San Francisco People Who Are Now Being Sheltered in Oakland" included the names and addresses of the people who had registered at the

chamber of commerce in Oakland, the "headquarters of the Oakland relief committee." Although this would later be listed as one of the official places to register in Oakland (to be on the centralized register), the advertisement in the newspaper directed that this location was the "Business Men's Headquarters in Oakland," where businessmen could "register and obtain information." The *Bulletin* printed another "official" register, "Where to Find Your Missing Friends" from Berkeley, which included "the names of all refugees who have sought shelter at the [Berkeley relief] committee headquarters." This role included names and new address if available, and "interspersed with these names and addresses are to be found occasional queries and requests for person to call at certain places or make their whereabouts known." The *Bulletin* also advertised "Lost Relatives Scan This Column," where it listed "notices ... handed to the Oakland office of *The Bulletin*." Besides the *Bulletin*'s own private newspaper register, it advertised an office for the *Portland Oregonian* for "Oregon, Washington, and Idaho survivors," and that "St. Louis Post-Dispatch requests all St. Louis people ... to register at the post office of the *Oakland Herald* immediately" such that "the names will be telegraphed at once to St. Louis." Lastly, there were club or fraternal organization registries. The *Bulletin* printed a list of names under the declaration that "a registration bulletin has been established at Odd Fellows' Hall, and to noon the following names had enrolled." *San Francisco Bulletin*, April 21, 1906.

96. For example, the *Bulletin* devoted over 2.5 pages to the list of persons trying to locate each other on April 21. By April 23, the list filled only 1.5 pages of the *Bulletin*. Quinn, *San Francisco Press and the 1906 Fire*, 5:26.

  The *San Francisco Call*, which was 6 or 8 pages in length from April 21 to 23, devoted nearly a page to listing those who had registered with the paper, listing information about people in hospitals and personal advertisements in the classified section on April 21. On April 22, the *Call* devoted even more space to giving information about people who were looking for others—almost a page. Another half page of personals indicates that people must have thought the newspaper would be an extraordinarily valuable place in which to put information for broadcast. By April 23, the *Call* had only one page of personals, April 24 and 25 saw the space decrease, and advertising came to dominate half the newspaper.

97. "Information Bureau Moves Headquarters," *San Francisco Chronicle*, May 11, 1906.

98. "Herald's Lists Reunite Many—Refugees and Friends Find Bureau Serviceable—Trained Men Are Bending Their Energies towards Bringing Loved Ones Together in These Days of Confusion," *Los Angeles Herald*, April 24, 1906. Although the *Herald* seemed to be quite obviously promoting itself, it also published the names and whereabouts of San Franciscans in Los Angeles so family members might be able to find them as well as "Want Information" ads for people who were sought. The *Herald*'s registration bureau "ended the terrible suspense of hundreds [of] those who,

since the awful catastrophe of Wednesday, have been anxiously waiting for news of dear ones."

99. "Herald Bureau Reunites Scores, Hundreds of Refugees Are Registered—Anxious Men and Women Delighted to Find Friends and Relatives by Applying to Newspaper Bureau," *Los Angeles Herald*, April 25, 1906. In a remarkably progressive move (San Francisco newspapers frequently directed Chinese or Japanese Californians to the embassy of their respective countries), the *Herald* said, "All nationalities are represented on the lists in the directory department, and scores of anxious men, women and children have been made happy through the ability of *The Herald* to inform them not only of the safety of loved ones but also their whereabouts." The *Herald* reported increasing requests from April 25 to 26, and then on April 27 noted "a marked decrease in the number of inquiries." This trend continued in the reporting of April 28, but the paper said, "There are still many who are anxious to hear of missing friends. ... Until requests for assistance cease *The Herald* will continue to publish lists daily." "Herald Bureau Keeps up Work—Refugee Reunions Brought about Daily—Daily Registration Are Made in Order to Establish Communication between Those Separated by the Earthquake," *Los Angeles Herald*, April 29, 1906.

100. "Bureau Locates Parted Friends—Herald Registration Lists Prove Effective—Through Publication of Names Many Happy Reunions Are Brought about among Those Separated," *Los Angeles Herald*, April 27, 1906.

101. Davies, *Saving San Francisco*, 73–74.

102. A multiplicity of registration bureaus sprang into action: "This work was under the special supervision of the Red Cross, and is expected to be in thorough working order to-day." "Hope Springs Eternal—About Registration and Information Bureaus," *San Francisco Chronicle*, April 21, 1906. An image from Golden Gate Park makes it clear that at least one registration was colocated with Red Cross tents and shows opportunities for free transportation on the Southern Pacific. Miles Bros, "Information Bureau, Refugee Camp, Golden Gate Park, No. 32," 1906, BANC PIC 1994.022—ALB v.1:02, Bancroft Library, University of California at Berkeley, http://ark.cdlib.org/ark:/13030/hb3g500744. Some sources note that the Red Cross and military set up a place for people to register, and it was widely used. "The Red Cross Association established an information bureau immediately after the earthquake, where people were requested to leave their names and addresses; at the present time 120,000 names have been recorded." Lucy B. Fisher, "A Nurse's Earthquake Experience," *American Journal of Nursing* 7, no. 2 (November 1906): 93; Harold French, "How the Red Cross Society Systematized Relief Work in San Francisco," *Overland Monthly* 48, no. 4 (October 1906): 203.

103. The official registration bureau in San Francisco was set up in the police headquarters and initially located on Fillmore Street (or in the mayor's office on Fillmore and Bush; different names and functions existed for this temporary building). Other

registration bureaus were in unburned areas of the city where the homeless were gathering.

104. The *Call* registry may have turned into one of the branch registration offices of the official registry, as the location of the original *Call* registration bureau was subsequently listed as a branch location of the official registration bureau (and it also stopped advertising registrants as its own). "Thousands of Helping Hands Relieve Plight of City," *San Francisco Call*, April 23, 1906. "The congestion at the information bureau at Police Headquarters on Fillmore street, near Bush, is being relieved by the following branch information bureaus: Lombard street and Presidio entrance, Baker and Fell streets, Jefferson square, Twenty-fifth and Mission streets. Those who have lost friends or homes in the city should register at the nearest bureau." See also the stereoptic card "A Temporary Relief Camp, Police Headquarters, and Registration Bureau on Van Ness Ave., San Francisco," 1906, BANC PIC 1989.018:35—STER, Bancroft Library, University of California at Berkeley, http://ark.cdlib.org/ark:/13030/hb7m3nb65c. The information bureaus located in Oakland were frequently cited as the locations to register, as many people had fled the fire by taking boats across the bay. The registration stations in Oakland were at city institutions such as the chamber of commerce. Registration bureau outposts were also in Alameda and Berkeley at the relief centers. "Registrations at Local Headquarters," *Sausalito News*, April 28, 1906.

105. "Registry Bureaus throughout City—Persons Seeking or in Possession of Information Should Make Themselves Known," *San Francisco Call*, April 23, 1906.

106. "How Refugees Find Friends," clippings, Rene Bine Collection, California Historical Society, San Francisco.

107. Ibid.

108. The report was published in various forms in a number of newspapers. One newspaper, for instance, said that Greely was "evidently depending on the newspapers himself. He says there are 300,000 persons homeless, that the city covers an area of twenty-five miles with no means of getting about except by walking, and he suggests that telegrams of inquiry be sent to the *Call*, *Chronicle* or *Examiner*." "New Relief Plan Adopted," *New York Sun*, April 26, 1906.

109. The Chinese, in the shadow of racist "exclusion" policies, had developed a robust Chinese-language press. The Chinese-language newspaper *Chung Sai Yat Po* (*China West Daily*) was instrumental in supporting and even motivating the Chinese American community as it contested the Progressive elite in San Francisco who wished to "move" Chinatown. That the Chinese press existed at all is, however, remarkable. The harrowing stories of the English-language press fleeing San Francisco to print newspapers in Oakland seems relatively simple considering that the English character set was available. The Chinese newspapers were handwritten for many months after the earthquake while new character sets were ordered. This is

perhaps yet another example that it is not the technology that made the press. Yumei Sun, "From Isolation to Participation: *Chung Sai Yat Po* and San Francisco's Chinatown, 1900–1920" (PhD diss., University of Maryland at College Park, 1999). See also Erica Y. Z. Pan, *The Impact of the 1906 Earthquake on San Francisco's Chinatown* (New York: P. Lang, 1995).

110. The April 28 issue said, "Many people survived the earthquake and fire and they're temporarily housed in Berkeley, so friends and relatives can come to Fook Wah 1561 Spring St Berkeley to look at the bulletin or to leave a message." The April 26 issue explained that people might find out about friends in Berkeley, but also included other messages: "Chinese women and children may move to 13th Street at the lakefront where there is a tall building for drink, food, and shelter; Chinese males may take up temporary lodgings on 13th Street at the lakefront in the tents that will become available tomorrow; Chinese may freely enter the city but may not congregate and give rise to disease. When it comes to pestilence, it is every man for himself." (Many thanks to Elisa Oreglia for help with translation.) For more on the experiences of Chinese Americans, see Pan, *Impact of the 1906 Earthquake on San Francisco's Chinatown*.

111. Rebecca Solnit, *A Paradise Built in Hell: The Extraordinary Communities That Arise in Disaster* (New York: Viking, 2009).

112. This is something that C. A. Bayly underscored in his discussion of how the British distorted Indian publics in their attempts to dominate them. Ricardo Roque, "Review of 'Empire and Information: Intelligence Gathering and Social Communication in India, 1780–1870,'" *Reviews in History* (2016); C. A. Bayly, *Empire and Information: Intelligence Gathering and Social Communication in India, 1780–1870* (Cambridge: Cambridge University Press, 1996).

113. On legibility, see James C. Scott, *Seeing Like a State: How Certain Schemes to Improve the Human Condition Have Failed* (New Haven, CT: Yale University Press, 1998).

114. "Devine and Red Cross to Distribute Supplies," *San Francisco Bulletin*, April 26, 1906.

115. "New Plans to Supply Food," *San Francisco Bulletin*, April 28, 1906.

116. "Alleged Waste at One Relief Station," *San Francisco Chronicle*, April 28, 1906.

117. Jones, *American Red Cross from Clara Barton to the New Deal*, 119; Davies, *Saving San Francisco*, 55–56.

118. Jones, *American Red Cross from Clara Barton to the New Deal*, 119.

119. Ibid., 117–119; Kevin Rozario, "Nature's Evil Dreams: Disaster and America, 1871–1906" (PhD diss., Yale University, 1996), 44.

120. "To Count People of San Francisco," *San Francisco Chronicle*, May 22, 1900.

121. "The climate of San Francisco is such that people do not really need a home. ... They move from lodging-house to lodging-house, wherever they can find credit, and it is difficult in the extreme to get track of them. ... They dodge the enumerator. Many of·them have criminal records and have good reasons for not wishing to be located. We generally argue them into giving the information we want. Sometimes, in stubborn cases, we have to get the policeman of that beat to show his star there and assure the man that the question must be answered. We do not call on the police except when absolutely necessary, because we cannot be so sure of the truth of the answers when they are thus." "Census Returns Coming in Well, Greatest Trouble in Lodging-House; Sometimes the Police Are Called in," *San Francisco Chronicle*, June 8, 1900.

122. "The registration was entrusted, as early as April 25, to Dr. C. C. Plehn of the University of California, and was completed in the second week of May." Edward T. Devine, "The Situation in San Francisco," *Charities and the Commons: A Weekly Journal of Philanthropy and Social Advance* 16, no. 9. (June 2, 1906): 304.

"Dr. Devine, immediately after his arrival, established a Department of Registration in order to facilitate the work of distribution. Aided by one hundred and fifty college students and professors, as well as teachers of the local department, under Professor C. C. Plehn of the University of California, the number of those who required aid was rapidly ascertained." French, "How the Red Cross Society Systematized Relief Work," 203.

The system for registration in San Francisco after the earthquake apparently took several days to develop, as Plehn apparently "had not finally perfected his scheme" on April 29. By May 7, it was reported that "in order to bring system and harmony into the work a uniform registration has been inaugurated throughout the city, and a large corps of school teachers is giving valuable aid to the workers at the relief stations." "Professor Plehn Plans the Registration Bureau," *San Francisco Call*, April 29, 1906; "Excellent Progress in the Organization of Relief Stations, Uniform Registration Through," *San Francisco Call*, May 7, 1906.

123. Letter from William W. Morrow to Charles L. Magee, secretary, American Red Cross, May 12, 1906: "We think a system [for distributing food] has been adopted that will make the distribution as nearly perfect as possible, and as the subject may be of some interest to the National Society, I enclose herewith the plan of registering of persons desiring food, the directions of registering applicants at relief station; also a registration card and a food card. ... You may, perhaps, find it interesting, and I would suggest that you show it to Mr. President Taft. The plan was devised by Professor C. C. Plehn of our State University, and we think it would be well to have it made a matter of record for future reference. The plan goes into effect immediately." "California Relief," *Red Cross Bulletin* 3 (July 1906): 19.

124. "The Plan for Registration," Record Group 200, Records of the American National Red Cross, 1881–1916, box 55, folder 815.6, "California, San Francisco

Earthquake and Fire 4/18/06—Relief other than Health," 1, National Archives, College Park, MD.

125. Ibid., 1. The version formally printed in the *Bulletin of the Red Cross* clarified that there were three types of people who could fill out cards: "(1) officers of the Relief Station; (2) workers of the Associated Charities; (3) representatives of the Central Registration Bureau." "Instructions for Registering Applicants at Relief Stations," *Red Cross Bulletin* 3 (July 1906): 23.

126. "Plan for Registration," 2.

127. "Instructions for Registering Applicants at Relief Stations," 24. Photographs show signs in front of relief tents "In G.G. Park [i.e., Golden Gate Park], 1906" (relief office tent), 1906, Bancroft Library, University of California at Berkeley, http://ark .cdlib.org/ark:/13030/hb438nb3tj.

128. Jones, *American Red Cross from Clara Barton to the New Deal*, 124–125.

129. Solnit, *Paradise Built in Hell*; Davies, *Saving San Francisco*, 46–48, 63–69.

130. Mary Roberts Smith, "Relief Work in Its Social Bearings," *Charities and the Commons: A Weekly Journal of Philanthropy and Social Advance* 16, no. 9. (June 2, 1906): 308–310.

131. Ibid.

132. A second registration process in July 1906 enabled the relief model based on the idea of "rehabilitation." *San Francisco Relief Survey: The Organization and Methods of Relief Used after the Earthquake and Fire of April 18, 1906* (New York: Russell Sage Foundation, 1913), 49.

Rehabilitation was underpinned by the philosophy of restoring people to their preearthquake circumstances, provided that those circumstances fit into the Progressive understanding of what middle-class life looked like. Davies, *Saving San Francisco*, 53–56.

133. Jones, *The American Red Cross from Clara Barton to the New Deal*, 124, 127–129.

Physician Margaret Mahoney wrote in a pamphlet, "It never semed [sic] to occur to those who undertook the function of distribution that it was their duty to get the goods to the people ... They could have opened many stations where supplies could be easily served, or better still, each camp might have been outfitted." Margaret Mahoney, pamphlet, *The Earthquake, the Fire, the Relief*, July 28, 1906. Mahoney's pamphlet is often quoted as an example of a critic who herself had the money to print her own pamphlet criticizing the Progressive era relief system. Mahoney's sentiments were similar to those of Mary Kelly, the working-class leader of the United Refugees. Davies, *Saving San Francisco*, 81; Jones, *American Red Cross from Clara Barton to the New Deal*, 124.

134. Adolphus W. Greely, "Army Report by Maj. Gen. A. W. Greely," RG200 National Archive Gift Collection, Records of the American National Red Cross, 1881–1916, box 54, folder 815.02, "California, San Francisco Earthquake and Fire 4/18/1906" (n.d., but written after 1908), 31.

135. Mahoney, *The Earthquake, the Fire, the Relief*. Furthermore, Mahoney described the process of filling out registration cards: "The only way to obtain supplies was to fill out cards containing humiliating and impertinent questions."

136. Smith, "Relief Work in Its Social Bearings," 308–310; Jones, *American Red Cross from Clara Barton to the New Deal*, 126.

137. Jones, in her analysis of the American Red Cross activities, described how the strategies to make the act of receiving aid as uncomfortable as possible succeeded as many no longer sought food rations. Jones, *American Red Cross from Clara Barton to the New Deal*.

138. This quasi-socialist group was run by women who resisted the manner in which aid was distributed, reasoning that the millions of dollars donated to victims should be split equally among the victims rather than having Progressive elites decide who was worthy. Davies documents how lawyers within the United Refugees sought to create an alternate order by encouraging refugees to pool their resources using a series of contracts and forms. Although the United Refugees failed to gain traction for its movement, it and other groups resisting the order established by the relief committee did prompt visits from the American National Red Cross and Massachusetts Red Cross. These groups presumably shared an ideological orientation with the relief committee, and thus found that there had been no wrongdoing by the relief committee. Davies, *Saving San Francisco*, 75–83; Jones, *American Red Cross from Clara Barton to the New Deal*, 127–129.

139. Davies, *Saving San Francisco*, 75–83.

140. The United Refugees attempted to set up an alternate public information infrastructure, but ultimately failed. Ibid.

141. Ibid., 82.

142. Letter from William W. Morrow to Charles L. Magee, 19.

143. Greely, "Army Report by Maj. Gen. A. W. Greely," 58–59. A fire in a suburb of Boston in 1908 destroyed half the city. Greely was sent as the governor's representative. "Then he appointed a committee to manage the relief work, the principal member being largely identical with those on the California relief committee. So far as possible they adopted as their standard the methods and even the phraseology of the San Francisco relief givers."

144. "Hope Springs Eternal—About Registration and Information Bureaus"; "Registry Bureaus throughout City—Persons Seeking or in Possession of Information Should Make Themselves Known."

145. "Five Hundred Die in Great Disaster, Search of the Ruins Begun, Facts Will Never Be Known as Many Lie in Unnamed Graves; Health Office—500 Bodies; The First Systematic Search for Bodies Started Yesterday," *San Francisco Chronicle*, April 23, 1906.

146. "Death List Placed at a Thousand, Bodies Taken from Graves, Spectators to Handle Remains Removed from North Beach Trenches," *San Francisco Chronicle*, April 25, 1906; Gladys Hansen, "Who Perished? A List of Persons Who Died as a Result of the Great Earthquake and Fire in San Francisco on April 18, 1906," Archives Publication No. 2, 1980, San Francisco Archives, San Francisco, 2–3.

147. "Send Lists of Killed and Wounded," *San Francisco Chronicle*, April 27, 1906; "List of Dead Is Increasing," *San Francisco Call*, April 28, 1906.

148. "Greely's 'Death Roll' was only additions to the lists of dead—this was the sixth list of dead." "Greely's Death Roll," *Washington Post*, April 28, 1906.

149. "Captain Winn of General Greely's staff, who has charge of the work, last evening reported that less than 800 bodies have thus far been recovered and that 290 seriously injured are in hospitals." "List of Dead is Increasing."

150. Major General Adolphus W. Greely, *Special Report on the Relief Operations Conducted by the Military Authorities of the United States at San Francisco* (Washington, DC: US Government Printing Office, 1906), 95.

151. Committee of Forty on Reconstruction, *Report [of] the Sub-Committee on Statistics to the Chairman and Committee on Reconstruction*, 1907, Bancroft Library, University of California at Berkeley, http://ark.cdlib.org/ark:/13030/hb996nb6vc.

152. Ibid.

153. Tom Graham, "Sunday Interview: Gladys Hansen," *San Francisco Chronicle*, April 14, 1996.

154. Hansen, "Who Perished?"

155. Ibid.

156. Gladys Hansen and Frank R. Quinn, "The 1906 'Numbers' Game" (paper of the San Francisco Earthquake Research Project, San Francisco, 1985).

157. "1906 List of Dead and Survivors," *San Francisco Virtual Museum*, July 22, 2010, http://www.sfmuseum.org/hist11/list.html.

158. "Hope Springs Eternal—About Registration and Information Bureaus"; "Registry Bureaus throughout City—Persons Seeking or in Possession of Information Should Make Themselves Known."

159. Hansen and Condon, *Denial of Disaster*. Scholars have argued that the economic impact of the insurance payments from the fire damage was so great that it caused a depression in the United States in 1907. Efforts to mitigate damage to the Bay Area economy, however, downplayed the earthquake's impact. Kerry A. Odell and Marc D. Weidenmeir, "Real Shock, Monetary Aftershock: The 1906 San Francisco Earthquake and the Panic of 1907," *Journal of Economic History* 64, no. 4 (December 2004): 1002–1027.

160. An initial draft of the Lawson Report, the postearthquake analysis published by the Carnegie Foundation, hints at some of the resistance from the business community that the seismologists encountered during their investigations for the report.

161. Estimates of the damages from the earthquake and fire range from $350 to $500 million in 1906 dollars; about 80 percent of the $229 million held in insurance policies was paid out. Odell and Weidenmeir, "Real Shock, Monetary Aftershock"; Gordon Thomas and Max Morgan Witts, *The San Francisco Earthquake* (New York: Dell Publishing, 1971).

162. Sara E. Wermiel, *The Fireproof Building: Technology and Public Safety in the Nineteenth-Century American City* (Baltimore: Johns Hopkins University Press, 2000); Scott Gabriel Knowles, *The Disaster Experts*: Mastering Risk in Modern America (Philadelphia: University of Pennsylvania Press, 2011).

163. Geschwind, *California Earthquakes*, 20.

164. "To Correct False Reports," *Oakland Herald*, May 2, 1906. "The realty men throughout the state are to make an effort to set the Easterners right so far as the results of the recent earthquake are concerned." The letter is signed by "executive secretary" Herbert Burdette of Los Angeles, and includes a sample six-paragraph letter that tries to minimize the damage and pinpoint most of it in "the Chinese quarter, which were a discredit."

165. Rozario, *Culture of Calamity*, 23. Furthermore, Rozario says that accounts that focus on the feelings of euphoria after the earthquake were primarily those of the middle class, which had various types of safety nets.

166. Hansen and Condon, *Denial of Disaster*.

167. Hansen and Quinn, "1906 'Numbers' Game," 4.

168. Drew Gilpin Faust, *This Republic of Suffering: Death and the American Civil War* (New York: Knopf, 2008).

Chapter 4

1. Kimberly Kay Massey, "A Qualitative Analysis of the Uses and Gratifications Concept of Audience Activity: An Examination of Media Consumption Diaries

Maintained during the 1989 Loma Prieta Earthquake" (PhD diss., University of Utah, 1993), 145.

The twenty-nine student accounts of media use during the earthquake describe a variety of experiences with the mass media. When communication studies doctoral student Massey analyzed and included these diaries in her 1993 dissertation, she found that the students were active and critical participants in media consumption during the disaster. Ibid., v.

2. Ibid., 146.

3. Ibid., 148.

4. Thomas M. Brocher, Robert A. Page, Peter H. Stauffer, and James W. Hendley II, *Progress toward a Safer Future since the 1989 Loma Prieta Earthquake*, Fact Sheet 2014–3092, US Geological Survey, 2014; "October 17, 1989 Loma Prieta Earthquake," US Geological Survey Earthquake Hazards Program, June 16, 2016, http://earthquake.usgs.gov/regional/nca/1989.

5. Scott Gabriel Knowles, *The Disaster Experts*: Mastering Risk in Modern America (Philadelphia: University of Pennsylvania Press, 2011), 163–166; Guy Oakes, *The Imaginary War: Civil Defense and American Cold War Culture* (New York: Oxford University Press, 1994); Tracy C. Davis, *Stages of Emergency: Cold War Nuclear Civil Defense* (Durham, NC: Duke University Press, 2007).

6. Ted Steinberg, "Smoke and Mirrors: The San Francisco Earthquake and Seismic Denial," in *American Disasters*, ed. Steven Biel (New York: NYU Press, 2001), 103–128; Carl-Henry Geschwind, *California Earthquakes: Science, Risk, and the Politics of Hazard Mitigation* (Baltimore: Johns Hopkins University Press, 2001).

7. In the years following the earthquake, many different government organizations published reports about the disaster. Although these reports were not created in the immediate earthquake aftermath, when people were making sense of what had happened, they still serve as evidence of the government's role in producing a historical narrative about the disaster. Academic work analyzing the mainstream media (television, radio, and newspapers) augments this evidence. In this chapter, I use the work of social scientists who studied the disaster immediately after it occurred—looking to both their writings in scholarly peer-reviewed venues and, more frequently, various reports published by the government. It is a tricky line to walk—to consider these scholarly accounts both as "sources" with their own disciplinary, methodological, and temporal biases as well as build on their analyses. This chapter is unique in its synthesis of many different kinds of sources to create an account of the postearthquake information order.

8. This is a form of the "conduit metaphor," and envisions it working unidirectionally. Michael Reddy, "The Conduit Metaphor: A Case of Frame Conflict in Our Language about Language," in *Metaphor and Thought*, ed. Andrew Ortony (Cambridge: Cambridge University Press, 1979), 284–324.

9. C. A. Bayly, *Empire and Information*: Intelligence Gathering and Social Communication in India, 1780–1870 (Cambridge: Cambridge University Press, 1996).

10. Several earthquake-planning guidelines from the 1980s provide definitions of *public information*. For example, the Bay Area Regional Earthquake Preparedness Project defined the function of public information as "continuous communications with the public through all available media to provide hazard warnings, official instructions and announcements, status of critical lifeline and emergency services, and damage information. This includes the operation of an emergency information center as part of an EOC [Emergency Operations Center], as well as provisions for meeting the needs of the press and public inquiries." Bay Area Regional Earthquake Preparedness Project, *County Comprehensive Earthquake Preparedness Planning Guidelines*, BAREPP 85–88, 1985, 29.

This definition of public information appears almost verbatim in a 1985 FEMA publication. Federal Emergency Management Agency and Southern California Earthquake Preparedness Project, *Comprehensive Earthquake Preparedness Planning Guidelines: City*, Earthquake Hazards Reduction Series 2, FEMA 73 (Washington, DC: Federal Emergency Management Agency, 1985), 54.

The 1988 FEMA *Workbook for State Governments* recommends that "the emergency public information (EPI)" section of a disaster plan include provisions for "an authoritative source for public information in an emergency" and the "use of a Joint Information Center (JIC) as a central location to coordinate the release of Emergency Public Information materials in an emergency." The workbook goes on to define *emergency public information* as "information which is disseminated primarily in anticipation of an emergency or at the actual time of an emergency and in addition to providing information as such, frequently directs action, instructs, and transmits direct orders." *Integrated Emergency Management System: Capability Assessment and Multi-Year Development Plan for State Governments—Workbook for State Governments*, Civil Preparedness Guide 1–36 (Washington, DC: Federal Emergency Management Agency, 1988), 3–39, A2.

11. *Multihazard Functional Plan* (San Francisco: City and County of San Francisco Office of Emergency Services, 1988), enclosure 1–5, 5.

12. The Plan is: Federal Emergency Management Agency, *Plan for Federal Response to a Catastrophic Earthquake*, 1987–720–298/60105 (Washington, DC: US Government Printing Office, 1987).

Although there were many federal disaster response plans in development at the time of the Loma Prieta earthquake, I look at this particular plan because it was actually activated by FEMA during the earthquake, though on a "limited" basis. State/ Federal Hazard Mitigation Survey Team, *Hazard Mitigation Opportunities for California: State and Federal Hazard Mitigation Survey Team Report for the October 17, 1989 Loma Prieta Earthquake, California*, FEMA-845-DR-CA (California Offices of Emergency Services and Federal Emergency Management Agency, 1990), 21.

13. For a description of the role of Public Affairs, see Federal Emergency Management Agency, *Plan for Federal Response to a Catastrophic Earthquake*, A16–17.

14. Ibid., A13, A16–17.

"The objective of the JIC is to coordinate all information to avoid duplication, confusion and conflicting information that often is produced in a crisis environment. To establish a one-stop news and information center for the media, the FEMA PIO must function as a manager to facilitate the rapid flow of accurate facts and figures." Federal Emergency Management Agency, *When Disaster Strikes: A Handbook for the Media* (Washington, DC: US Government Printing Office, 1985), 13.

Another relevant area, Emergency Support Function #5, the Damage Information Annex, addressed how officials should "gather, collate, and disseminate information on damages (including casualties) following a catastrophic earthquake" for other disaster responders; this might inform public information, but it was not intended for the public. Federal Emergency Management Agency, *Plan for Federal Response to a Catastrophic Earthquake.*, 5–1. This annex includes templates for how to create damage information records.

15. Federal Emergency Management Agency, *Plan for Federal Response to a Catastrophic Earthquake*, xii–xiii; State/Federal Hazard Mitigation Survey Team, *Hazard Mitigation Opportunities for California*, 21.

16. Early reports did not indicate widespread devastation, and so "the decision was made to follow normal emergency operating procedures, rather than activate the recently completed and tested *Draft Plan for a Catastrophic Earthquake in the San Francisco Bay Region*. The State/Federal Hazard Mitigation Survey Team, *Hazard Mitigation Opportunities for California*, 21.

17. Ibid., 1.

18. *Multihazard Functional Planning Guidance* (Sacramento: State of California Governor's Office of Emergency Services, 1985), vii.

19. Ibid., 153, Enclosure A-6.

20. Ibid., 155, Enclosure A-6.

21. *Multihazard Functional Planning Guidance* (Sacramento: State of California Governor's Office of Emergency Services, 1985), 180–184, attachment A-6-C, exhibit 7.

22. Ibid., 159–162, attachment A-6-A, enclosure A-6.

The radio messages are for different scenarios: when there is "no information available"; an "update on earthquake," which gives information on deaths, homes damaged, magnitude, area affected, and epicenter; and a "summary statement for the media." Ibid., 217–220, attachment A-1-A.

23. Different plan sections were written and released at various dates in 1989, but all before the October 17, 1989 earthquake. Thomas M. Fante, *State Emergency Plan* (Sacramento: State of California Governor's Office of Emergency Services, 1989).

24. The *State Emergency Plan* suggests a more restricted view of how the Public Information Officers should communicate with the public than that in the *Multihazard Functional Planning Guidance*. It doesn't include radio scripts for communicating directly with the public or make any mention of setting up a Joint Information Center, both included in federal assessment guidelines. According to the *State Emergency Plan*, the job of a public information officer is to "enlist the cooperation of local, statewide, national, and international media in relaying emergency guidance to the affected public and providing status information to their audiences." Furthermore, communicating with the public is restricted: "Public information officers use the telephone to inform the media and, primarily at the local level, to respond to inquiries from the public." Ibid., annex L, L2, L7.

25. *Multihazard Functional Plan*, 315–317, Attachment A-6-C, Enclosure A-6.

26. Ibid., 82, Enclosure 1-5.

27. Ibid., 311, Attachment A-6-A, Enclosure A6.

28. "East Bay to Rehearse for Earthquake," *San Francisco Chronicle*, August 7, 1989; Ted Bell, "Disaster Drill Gets a Jolt of Reality," *Sacramento Bee*, August 9, 1989; Dave O'Brian, "Quake Prompts California to Come to Grips with Its Faults," *San Jose Mercury News*, August 9, 1989.

29. "Teams of specialists, ranging from procurement experts to construction managers, huddled quietly around computers, printers and TV screens in a sprawling building on the south side of Sacramento, shuttling notes and memos to each other. Others marked up chalkboards with the latest county-by-county victim totals, products of a computer's imagination. ... Reporters were cautioned not to bother the 'players' with questions, but some paid no heed, much to the irritation of the organizers." "Earthquakes on Paper Help with the Real Ones," *San Jose Mercury News*, August 9, 1989.

30. Bell, "Disaster Drill Gets a Jolt of Reality."
    A 1991 report from the US Government Accounting Office (GAO) asserted that spending on preparedness activities such as RESPONSE-89 did help with the Loma Prieta response. There are a number of references to how RESPONSE-89 helped identify shortcomings and practice communication between various parties involved with disaster response. *Disaster Assistance: Federal, State, and Local Responses to Natural Disasters Need Improvement*, RCED-91-43 (Washington, DC: US General Accounting Office, 1991), http://www.gao.gov/products/RCED-91-43; William M. Brown III and Carl Mortenson, "Earth Science, Earthquake Response, and Hazard Mitigation: Lessons from the Loma Prieta Earthquake," in *The Loma Prieta, California, Earthquake of October 17, 1989—Recovery, Mitigation, and Reconstruction*, ed. Joanne M Nigg, US Geological Survey Professional Paper 1553-D (Washington, DC: US Government Printing Office, 1998), D81–90, http://pubs.usgs.gov/pp/pp1553/pp1553d.

31. Alexandra Hayne, city editor of the *Register-Pajaronian*, quoted in *Covering the Quake: A Transcript of a Symposium Held on December 9, 1989 on the Media's Coverage of the 1989 Bay Area Earthquake* (Berkeley: Graduate School of Journalism, University of California at Berkeley, 1990), 14.

32. Richard J. Rappaport, "The Media: Radio, Television, and Newspapers," in *The Loma Prieta, California, Earthquake of October 17, 1989—Lifelines*, ed. Ansel J. Schiff, US Geological Survey Professional Paper 1552-A (Washington, DC: US Government Printing Office, 1998), A45.

33. Eric Newton, assistant managing editor of the *Oakland Tribune*, quoted in *Covering the Quake*, 5; Rappaport, "Media: Radio, Television, and Newspapers," A45.

34. Conrad Smith, *Media and Apocalypse: News Coverage of the Yellowstone Forest Fires, Exxon Valdez Oil Spill, and Loma Prieta Earthquake* (Westport, CT: Greenwood Publishing Group, 1992). The proceedings of a conference about the media coverage of the earthquake called this the "San Francisco Syndrome." *Covering the Quake.*

35. "Transcription of Television News from Channel 3 and Other Sources," notes taken by Martha Savage, seismologist, typed by Rovert Sydnor, geologist; "Earthquake Notes," October 17, 1989, Loma Prieta Earthquake Files, Governor's Office of Planning and Research, Administration and Personnel Services, California State Archives, Sacramento.

36. There were 102 shots of the Marina, 69 shots of the freeway in Oakland, and 27 from Santa Cruz. In the days that followed, on the national television news programs there were 13 stories that focused on San Francisco, 10 on Oakland, and 5 on Santa Cruz. Smith, *Media and Apocalypse*, 127–129, 120.

37. Eric Newton, assistant managing editor of the *Oakland Tribune*, and Jane Gross, reporter for the *New York Times*, quoted in *Covering the Quake*, 11–12.

38. Everett M. Rogers, Matthew Berndt, John Harris, and John Minzer, "Accuracy in Mass Media Coverage," in *The Loma Prieta Earthquake: Studies of Short-Term Impacts*, ed. Robert C. Bolin (Boulder: Institute of Behavioral Science, University of Colorado, 1990), 46–47.

39. Richard Clay Wilson Jr., *The Loma Prieta Quake: What One City Learned* (Washington, DC: International City Management Association, 1991), 41.

40. The Red Cross tried to take the funds that were donated to Loma Prieta relief and add them to the general relief fund, but "in deference to donors' requests, the Red Cross determined that all donations designated for northern California would stay in northern California." However, and in a 1991 report about the earthquake, the Red Cross emphasized its stance "of encouraging donations to the national Disaster Relief Fund, rather than for specific disasters, to assure adequate response to all disasters." *Meeting the Loma Prieta Challenge: An Interim Report of the Northern*

California Earthquake Relief and Preparedness Project (Burlingame, CA: American Red Cross, 1991).

41. City of Watsonville, Office of the City Manager, report included in California Seismic Safety Commission, *Loma Prieta's Call to Action: Report on the Loma Prieta Earthquake of 1989* (Sacramento: California Seismic Safety Commission, 1991), 95.

42. National Research Council, "Overview: Lessons and Recommendations from the Committee for the Symposium on Practical Lessons from the Loma Prieta Earthquake," in *Practical Lessons from the Loma Prieta Earthquake: Report from a Symposium Sponsored by the Geotechnical Board and the Board on Natural Disasters of the National Research Council* (Washington, DC: National Academies Press, 1994), 4.

43. Kathleen Tierney, "Emergency Medical Care Aspects of the Loma Prieta Earthquake," University of Delaware Disaster Research Center, Preliminary Paper Series, no. 161 (1991): 7–8.

44. State/Federal Hazard Mitigation Survey Team, *Hazard Mitigation Opportunities for California*, 21–22.

45. All the plans direct the persons in charge of public information in the government disaster response organization (be it someone from public affairs or a public information officer) to work on rumor control.

46. Of course, people who experience a disaster do not necessarily have technical expertise about seismic safety, so they may not be in a position to provide instructions about what to do next or might not be able to see specific kinds of geologic hazards.

47. This was particularly ironic because the USGS seemed to fill many public communication roles after Loma Prieta: it systematically gathered data about the earthquake, and met daily to share that information with each other and the public. Brown III and Mortenson, "Earth Science, Earthquake Response, and Hazard Mitigation."

48. State/Federal Hazard Mitigation Survey Team, *Hazard Mitigation Opportunities for California*, 25.

49. Louise K. Comfort, *Shared Risk: Complex Systems in Seismic Response* (New York: Pergamon Press, 1999), 172; Louise K. Comfort, "Reconnaissance Report: The Loma Prieta Earthquake, October 17, 1989: Communication and Coordination in Emergency Response and Recovery," submitted to Robert Olson, Team Leader, Social Sciences, Reconnaissance Team, Loma Prieta Earthquake of October 17, 1989 (January 2, 1990); Institute of Governmental Studies Library, University of California at Berkeley.

According to the findings of the Earthquake Engineering Research Institute's *Reconnaissance Report*, which Comfort helped author, all jurisdictional levels of disaster response (federal, state, and local) are responsible for "communication of information [to the public] regarding the event and restatement of the goal of disaster operations," but there was also the opportunity for the public to inform the disaster response professionals in assessing the "needs and status of the affected community." Robert A. Olson, Shirley Mattingly, Charles Scawthorn, Jelena Pantelic, Dennis Mileti, Colleen Fitzpatrick, Steven Helmericks, et al., "Socioeconomic Impacts and Emergency Response," *Earthquake Spectra* 6, no. S1 (May 1, 1990): 408.

50. Comfort, "Communication and Coordination in Emergency Response and Recovery," 7.

51. Ibid., 2–7; Olson et al., "Socioeconomic Impacts and Emergency Response," 409.

52. California Seismic Safety Commission, *Loma Prieta's Call to Action*, 21; Ansel J. Schiff, Alex Tang, Lawrence F. Wong, and Luis Cusa, "Communication Systems," in *The Loma Prieta, California, Earthquake of October 17, 1989—Lifelines*, ed. Ansel J. Schiff, US Geological Survey Professional Paper 1552-A (Washington, DC: US Government Printing Office, 1998), A24; State/Federal Hazard Mitigation Survey Team, *Hazard Mitigation Opportunities for California*, 25.

53. California Seismic Safety Commission, *Loma Prieta's Call to Action*, 21; State/Federal Hazard Mitigation Survey Team, *Hazard Mitigation Opportunities for California*, 25.

54. Schiff et al., "Communication Systems," A24–25.

55. Ibid., A27–28; Tierney, "Emergency Medical Care Aspects of the Loma Prieta Earthquake," 5.
    Although problems did exist with the telecommunication infrastructure, it is instructive to consider them in light of a preliminary engineering report: "In general, telephone systems performed better than expected, especially considering the seismic forces that equipment in the epicenter area had to withstand." This suggests that the issues that Comfort identified in her research were likely problems with organizational communication, exacerbated by the need to adopt work-arounds to telecommunication infrastructure. EQE Engineering, *The October 17, 1989 Loma Prieta Earthquake* (San Francisco: EQE Engineering, 1989), 24.

56. Eileen Cahill Maloney, speaking at the San Francisco departmental debriefing on Command Center Operations following the October 17, 1989 earthquake, San Francisco, November 1, 1989, http://www.sfmuseum.net/quake/debrief.html. Mahoney was the press secretary to the mayor; she would have fulfilled the public information officer role in San Francisco.

57. Deborah Acosta, town manager for Los Gatos, quoted in California Seismic Safety Commission, *Loma Prieta's Call to Action*, 19.

58. Olson et al., "Socioeconomic Impacts and Emergency Response," 407.

59. State/Federal Hazard Mitigation Survey Team, *Hazard Mitigation Opportunities for California*, 25.

60. Wilson, *The Loma Prieta Quake*, 33.

61. Ibid., 32.

62. *Covering the Quake*; Rogers et al., "Accuracy in Mass Media Coverage," 47.

63. Rogers et al., "Accuracy in Mass Media Coverage," 48, 50, 53.

64. Smith analyzed earthquake-related newspaper stories in the *New York Times, Washington Post, Los Angeles Times,* and *San Jose Mercury News* for six months after the earthquake as well as the national television news stations CBS, NBC, and ABC. Of the 240 sources used in multiple stories, 50.1 percent were government officials. Of the 2,256 unique sources cited in the newspaper stories analyzed, 36.3 percent were from government agencies, elected officials, or safety officials. Smith, *Media and Apocalypse*, 135–137.

65. Other radio systems were to help emergency responders communicate with each other. *Multihazard Functional Plan*, 259–273, enclosure A3; 263, enclosure A-3.

66. Federico Subervi-Vélez, Maria Denney, Anthony Ozuna, Clara Quintero, and Juan-Vicente Palerm, *Communicating with California's Spanish-Speaking Populations: Assessing the Role of the Spanish-Language Broadcast Media and Selected Agencies in Providing Emergency Services* (Berkeley: California Policy Seminar, University of California at Berkeley, 1992), 14, 47–48.

67. Maloney, speaking at the San Francisco departmental debriefing.

The Emergency Broadcasting System was used locally in San Francisco and Santa Cruz Counties to broadcast postearthquake announcements. KNBR broadcast Mayor Art Agnos's emergency speech as well as translations (provided by the mayor's aides) in Spanish, Chinese, Vietnamese, and Tagalog. Subervi-Vélez, et al., *Communicating with California's Spanish-Speaking Populations*, 51–52.

68. Rogers et al., "Accuracy in Mass Media Coverage," 48.

69. Wilson, *Loma Prieta Quake*, 35.

70. California Library Association Earthquake Relief Grant Ad Hoc Committee, *Earthquake Preparedness Manual for California Libraries* (Sacramento: California Library Association, 1990), 25 (emphasis in original).

71. Ibid., 25.

72. Ibid, 25.

73. Ibid, 25.

74. Wilson, *Loma Prieta Quake*, 35.

75. *City of Oakland, Loma Prieta Earthquake, after Action Report* (Oakland, CA: Oakland Office of Emergency Services, 1990), executive summary, http://www .sfmuseum.org/oakquake/1.a.html.

76. Art Jensen, speaking at the San Francisco departmental debriefing on Command Center Operations following the October 17, 1989 earthquake, San Francisco, November 1, 1989, http://www.sfmuseum.net/quake/debrief.html.

77. Tony Russomanno, reporter for KGO-TV, and Peter Laufer, reporter for KCBS, quoted in *Covering the Quake*, 2–3, 10.

78. "With little or no information coming from these official sources, reporters during the first hours after the earthquake turned to the various unofficial sources for a great deal of their information. ... In the absence of more concrete information, reporters used the accounts of eyewitnesses to create an imprecise image of the disaster's severity for their audience." Rogers et al., "Accuracy in Mass Media Coverage," 48.

79. Smith, *Media and Apocalypse*, 138.

80. I am borrowing this term from Star, and drawing inspiration from Steve Graham and Simon Martin. Susan Leigh Star, "Orphans of Infrastructure: A New Point of Departure" (paper presented at the Future of Computing: A Vision, Oxford Internet Institute, University of Oxford, March 2007); Simon Marvin and Stephen Graham, *Splintering Urbanism*: Networked Infrastructures, Technological Mobilities, and the Urban Condition (London: Routledge, 2001).

81. *Multihazard Functional Planning Guidance*, 162, attachment A6; Joe Tran and Dennis Conkin, "Refugees Again: Temblor Spurs Asian Flight from TL," *Tenderloin Times*, November 1989; Dennis Conkin and Sara Colm, "Temblor Confusion Brings S.E. Asian Campers to Civic Center," *Tenderloin Times*, October 20, 1989, earthquake extra edition; Johnny Ng, "Chinatown's Seniors Frightened by Quake," *Asian Week*, October 27, 1989; Johnny Ng, "Critics: FEMA Ignored Chinese Quake Victims," *Asian Week*, December 1, 1989; Subervi-Vélez et al., *Communicating with California's Spanish-Speaking Populations*.

82. Steven A. Chin, "City Turned into a Tower of Babel: Foreign-Speaking Residents Were Left with No Information," *San Francisco Examiner*, October 30, 1989.

83. Ibid.; Tran and Conkin, "Refugees Again"; Conkin and Colm, "Temblor Confusion Brings S.E. Asian Campers to Civic Center."

84. *Tenderloin Times, Earthquake Extra Edition* (San Francisco, 1989); Johnny Ng, "Hsieh, Officials Tell S.F. Chinatown How to Prepare for Next Quake," *Asian Week*, November 17, 1989; William Wong, "Editorial," *Asian Week*, November 3, 1989.

85. Historian Ted Steinberg has referred to "the federal disaster bureaucracy" as "the secret benefactor of the middle-class." Marian Moser Jones, *The American Red Cross*

from Clara Barton to the New Deal (Baltimore: Johns Hopkins University Press, 2013); Andrea Rees Davies, *Saving San Francisco*: Relief and Recovery after the 1906 Disaster (Philadelphia: Temple University Press, 2011); Ted Steinberg, *Acts of God*: The Unnatural History of Natural Disaster in America, 2nd ed. (New York: Oxford University Press, 2006), 178.

86. Megan Finn, Janaki Srinivasan, and Rajesh Veeraraghavan, "Seeing with Paper: Government Documents and Material Participation," in *2014 47th Hawaii International Conference on System Sciences* (Los Alamitos, CA: IEEE Computer Society, 2014), 1515–1524; Davies, *Saving San Francisco*.

87. The elderly who lived on a fixed income, those who lived in units but were not on leases, those who shared housing, those who lived in single-room occupancy units for short periods of time, and those at risk of being homeless by occupying some of the oldest and least well-kept buildings in the Bay Area were left homeless by the earthquake and unable to get aid. Of the 869 living units in 8 severely damaged single-room occupancy buildings in Oakland, only 363 individuals could provide the paperwork that qualified them for FEMA assistance. Advocates for the people affected by these FEMA policies eventually filed a lawsuit against FEMA, and in 1990, won $23.04 million to replace low-income housing (75 percent of which was to be paid by FEMA, and the rest by the state). *Disaster Assistance: Federal, State, and Local Responses to Natural Disasters Need Improvement*, 55–56; California Seismic Safety Commission, *Loma Prieta's Call to Action*, 55.

88. There was even a question of whether predisaster homeless should be in earthquake shelters. The director of Emergency Services of Oakland described the sheltering system as "a very middle-class system—never did anybody [think] that Mr. and Mrs. Jones would be residents of a Single-Room Occupancy hotel. Like I told one reporter, 'This is not "Ozzie and Harriet Go to the Shelter." These are some real borderline people with some big problems.' I'm the first to admit that these people had social problems while they were residents of these hotels, but they *were* residents of dwellings in Oakland that were destroyed by the earthquake, so why should they not qualify for the same benefits as anybody else?" Henry Renteria, quoted in Bay Area Regional Earthquake Preparedness Project, "When the Going Got Tough," *Networks: Earthquake Preparedness News* 5, no. 1 (Winter 1990): 17.

89. Kathleen Tierney, "Emergency Preparedness and Response," in *Practical Lessons from the Loma Prieta Earthquake: Report from a Symposium Sponsored by the Geotechnical Board and the Board on Natural Disasters of the National Research Council* (Washington, DC: National Academies Press, 1994), 112.

90. Mary C. Comerio, "Hazard Mitigation and Housing Recovery: Watsonville and San Francisco One Year Later," in *The Loma Prieta, California, Earthquake of October 17, 1989—Recovery, Mitigation, and Reconstruction*, ed. Joanne M. Nigg, US Geological Survey Professional Paper 1553-D (Washington, DC: US Government Printing Office, 1998), D31.

91. Catherine M. Simile, "Disaster Settings and Mobilization for Contentious Collective Action: Case Studies of Hurricane Hugo and the Loma Prieta Earthquake" (PhD diss., University of Delaware, 1995), 107.

92. Barbara Garcia, interview by Jennifer Jordan, June 12, 1990, in Irene Reti, ed., *The Loma Prieta Earthquake of October 17, 1989: A UCSC Student Oral History Documentary Project* (Santa Cruz: University of California at Santa Cruz Library, 2006), 21–22.

93. Simile, "Disaster Settings and Mobilization for Contentious Collective Action," 82–84.

94. Because Watsonville experienced such a great loss of housing, with little available slack, shelter was a major issue after the disaster. In some cases, the difficulties of emergency sheltering were compounded by the relationship between Latinos and the city of Watsonville, and the citizenship status of undocumented Latinos. Most people who could no longer live in their houses lived in sanctioned Red Cross shelters, where food and medical help were available. Initially, two unsanctioned parks became home for several hundred, mostly Latino refugees. According to many sources, some services and parks were preferable because people were afraid that federal or quasi-federal organizations would work with the Immigration and Naturalization Services. One of the parks for refugees became a Red Cross–run park on October 21 and 22; the other eventually had Red Cross food service. Some Latinos feared the Immigration and Naturalization Services because they were undocumented aliens; others feared it because they were documented but thought that they were not allowed to solicit welfare services, not realizing that emergency relief is granted to anyone regardless of citizenship or legal status. Subervi-Vélez et al., *Communicating with California's Spanish-Speaking Populations*, 61.

95. David Tuller, "Illegal Immigrants Can Get Aid," *San Francisco Chronicle*, October 24, 1989.

96. George Cathron, "Earthquake Night News: For English Speakers Only," *Bay Guardian*, November 1, 1989, quoted in Subervi-Vélez et al., *Communicating with California's Spanish-Speaking Populations*, 49.

97. Subervi-Vélez et al., *Communicating with California's Spanish-Speaking Populations*, 9–11.

98. For the whole population, on the day of the earthquake, 49.6 percent of people found that radio was the best source of information and 42.4 percent found that television was the best source of information. In the days following the earthquake, almost 60 percent of people found television to be the best source of information. For the Hispanic population, "after the earthquake … 67.6% of Hispanics found television to be the best source of information, 16.7% found radio, and 9.4% looked to newspapers [as] the best source of information." The data from this work came from the Survey Research Center at the Institute for Social Science research at the University of California at Los Angeles. The 656 structured interviews were collected

in 1990 using random phone numbers and had a 70 percent response rate. William E. Lovekamp, "Gender, Race/Ethnicity, and Social Class Differences in Disaster Preparedness, Risk, and Recovery in Three Earthquake-Stricken Communities" (PhD diss., Southern Illinois University at Carbondale, 2006), 155–156, 125, 140.

99. Subervi-Velez et al, *Communicating with California's Spanish-Speaking Populations*, 53.

100. Ibid., 43, 53, 56.

101. The Southern California Earthquake Preparedness Project, *The Coalinga Earthquake Initial Steps toward Recovery [DRAFT]*, May 25, 1983, 8, box 52, folder R118.109, Earthquake Studies—Earthquake Reports [Series], Records of the Division of Mines and Geology, Department of Conservation, California State Archives.

102. *Multihazard Functional Planning Guidance*, 162, attachment A-6-A; 171, attachment A-6-C; 191–195, attachment A-6-E.

103. The instruction that the public information function includes "ensuring emergency information is translated for special populations (non-English speaking, persons with disabilities)" appears in 1980s' Bay Area Regional Emergency Preparedness Project publications. Federal Emergency Management Agency and Southern California Earthquake Preparedness Project, *Comprehensive Earthquake Preparedness Planning Guidelines*, 55; Bay Area Regional Earthquake Preparedness Project, *County Comprehensive Earthquake Preparedness Planning Guidelines*, 29.

104. Steve Twomey, "Immigrants Afraid to Ask for Help; Many Hispanic Workers Have Seen Their World Collapse," *Washington Post*, October 23, 1989.

105. "We quickly started talking to every major Spanish-speaking media that we could get our hands on, and all the media that we could regarding the area. ... The Latino community was just basically ignored. Most of the politicians that came to town ignored the Latino community, because we were ignored by the public officials. We don't have any representation there, and they were not bringing people to come to look at the epicenter of the earthquake. So the city of Watsonville contributed to the lack of press coverage, as far as I was concerned." Garcia interview, quoted in Reti, *Loma Prieta Earthquake of October 17, 1989*, 25–26.

106. City of Watsonville, Office of the City Manager, quoted in California Seismic Safety Commission, *Loma Prieta's Call to Action*, 96.

107. Subervi-Vélez et al., *Communicating with California's Spanish-Speaking Populations*, 56.

108. Ibid., 52–56.

109. Frank Viviano, "How Relatives Overseas Kept Informed," *San Francisco Chronicle*, October 23, 1989; Subervi-Vélez et al., *Communicating with California's Spanish-Speaking Populations*, 54–55.

110. Brenda Phillips and Mindy Ephraim, "Living in the Aftermath: Blaming Processes in the Loma Prieta Earthquake," Natural Hazards Research Applications Information Center, Institute of Behavioral Science, University of Colorado, Natural Hazards Working Paper Series, no. 80 (September 1992), 10.

111. Subervi-Vélez et al., *Communicating with California's Spanish-Speaking Populations*, 62.

112. Simile, "Disaster Settings and Mobilization for Contentious Collective Action," 174–175.

113. Garcia interview, quoted in Reti, *Loma Prieta Earthquake of October 17, 1989*.

In addition to the support of groups such as Salud Para La Gente, people in the shelters organized and supported each other—even assisting with shelter management when allowed. Ruth M. Laird, "Ethnography of a Disaster" (master's thesis, San Francisco State University, 1991); Brenda Phillips, "Sheltering and Housing Low-Income and Minority Groups in Santa Cruz County after the Loma Prieta Earthquake," in *The Loma Prieta, California, Earthquake of October 17, 1989—Recovery, Mitigation, and Reconstruction*, ed. Joanne M. Nigg, US Geological Survey Professional Paper 1553-D (Washington, DC: US Government Printing Office, 1998), D24.

114. Based on an analysis of newspapers and interviews, a researcher wrote the following: "Our data suggest that this rally was, for many Latino residents, the first opportunity to learn of aid sources. The media indicated varying perceptions of Latino needs on the part of city officials. Community groups set up information tables at the rally, while the city did not provide any." Phillips and Ephraim, "Living in the Aftermath," 4.

115. The city had been sued and was ordered by Supreme Court judges to redistrict; after the redistricting, all the city council members lived in the same district. Garcia summed up the position of Salud Para La Gente within the Latino community: "We're a major player in this community in providing services to the Latino community. We are the power representation of it. We don't have it in the city council. We don't have it in the county. We don't have it anywhere. And so community-based organizations, Latino-based are the power brokers within the community." Garcia interview, quoted in Reti, *Loma Prieta Earthquake of October 17, 1989*, 35.

116. One of the authors of Watsonville's disaster plans acknowledged after the earthquake that groups such as Salud Para La Gente were integral to disaster response: "I only wish we had coordinated that prior to the disaster so we would have had a better handle on the capabilities our community had in dealing with those kinds of issues." Gary Smith et al., "Watsonville's Ready with New Disaster Plan," *Western City* 62, no. 8 (August 1986): 4–6; Gary Smith, Watsonville Fire Department chief and emergency services director, quoted in California Seismic Safety Commission, *Loma Prieta's Call to Action*, 18.

The city of Watsonville report also reflected the significance of collaboration with local groups: "The coordination of services between government and community service agencies is very important." The report went on to recommend the creation of community response groups that "need to include representatives who reflect the socioeconomic and cultural makeup of the community so that a variety of service needs are planned for." Ibid., 89, 94.

117. City of Watsonville, Office of the City Manager, quoted in California Seismic Safety Commission, *Loma Prieta's Call to Action*, 96.

118. Robert C. Bolin and Lois M. Stanford, "Emergency Sheltering and Housing of Earthquake Victims: The Case of Santa Cruz County," in *The Loma Prieta, California, Earthquake of October 17, 1989—Public Response*, ed. Patricia Bolton, US Geological Survey Professional Paper 1553-B (Washington, DC: US Government Printing Office, 1993), B46.

119. The Watsonville Disaster Assistance Center "was located on the outskirts of the city. ... The City had to organize a special transportation network from the Red Cross shelters to the center." City of Watsonville, Office of the City Manager, quoted in California Seismic Safety Commission, *Loma Prieta's Call to Action*, 91.

120. Tuller, "Illegal Immigrants Can Get Aid"; John H. Cushman Jr., "The California Quake; Tangle of Paperwork Snarls Path to Aid for Farm Workers and Shop Owners," *New York Times*, October 25, 1989.

121. According to one anthropologist, "Spanish language instructions and assistance had not been readily provided by City Hall or by FEMA or by the Red Cross in the first days. Signs, from 'red tags' on houses to traffic changes, directions to FEMA and assistance, were not posted in Spanish." Laird, "Ethnography of a Disaster," 58.

122. Phillips and Ephraim, "Living in the Aftermath," 11.

123. State/Federal Hazard Mitigation Survey Team, *Hazard Mitigation Opportunities for California*, 25.

124. "Mainstream definitions abounded in the eligibility components and in agencies' action to separate combined or extended families or households into separate nuclear families for processing." Laird, "Ethnography of a Disaster," 118; Bolin and Stanford, "Emergency Sheltering and Housing of Earthquake Victims," B47.

Chapter 5

1. Patrick Sheriff, ed., *2:46: Aftershocks: Stories from the Japan Earthquake* (London: Enhanced Editions Ltd., 2011).

2. As with the previous chapters, here I focus on the public information infrastructure. Though there are complex behind-the-scenes information infrastructures in

place for professional responders to share information, I include these particular infrastructures only when public communication is also included in some way.

3. When I use the term *bureaucratic technology*, I draw on a broad definition of technology—one that uses technology to refer to prescriptive techniques.

4. Axel Bruns and Jean Burgess, "Crisis Communication in Natural Disasters: The Queensland Floods and Christchurch Earthquakes," in *Twitter and Society*, ed. Katrin Weller, Axel Bruns, Jean Burgess, Merja Mahrt, and Cornelius Puschmann (New York: Peter Lang, 2014), 89:373–384.

5. The study of information infrastructure is often referred to as "boring" because infrastructure is not glamorous and tends to be hidden, and its implications are not obvious. Star frequently refers to her colleagues who study infrastructure as members of the "Society of People Interested in Boring Things." Susan Leigh Star, "The Ethnography of Infrastructure," American Behavioral Scientist 43, no. 3 (November 1, 1999): 377–391.

6. Later I use Tarleton Gillespie's terminology to describe how social media companies produce *calculated publics* in some sense. The state does this as well, though its publics are broader. Tarleton Gillespie, "The Relevance of Algorithms," in Media Technologies: Essays on Communication, Materiality, and Society, ed. Tarleton Gillespie, Pablo J. Boczkowski, and Kirsten A. Foot (Cambridge, MA: MIT Press, 2014), 167–194.

7. Paul DiMaggio and Walter W. Powell, "The Iron Cage Revisited: Collective Rationality and Institutional Isomorphism in Organizational Fields," *American Sociological Review* 48, no. 2 (1983): 147–160.

8. The Incident Command System requires that the government responders use "clear text," meaning that they use "common terminology" so that when people from different organizations get together, they all have the same names for the roles that organize response. *Foundation for the Standardized Emergency Management System (SEMS)* (Sacramento: California Emergency Management Agency, 2010), http://www.caloes.ca.gov/cal-oes-divisions/planning-preparedness/standardized-emergency-management-system.

9. *National Incident Management System* (Washington, DC: US Department of Homeland Security, 2008), 140.

10. Michael K. Buckland, "Information as Thing," *Journal of the American Society for Information Science* 42, no. 5 (1991): 351–360.

11. NIMS is not a disaster response plan but instead describes an incident management framework that codifies certain principles for disaster response. The analysis in this chapter focuses on the information and communication practices that are articulated in the 2008 (and most current as of this writing) version of NIMS.

The national standardization of disaster response plans was propagated through the civil defense programs of the 1950s and 1960s, which built out the long arm of federal disaster response planning and training apparatus. Scott Gabriel Knowles, *The Disaster Experts*: Mastering Risk in Modern America (Philadelphia: University of Pennsylvania Press, 2011); Kevin Rozario, *The Culture of Calamity*: Disaster and the Making of Modern America (Chicago: University of Chicago Press, 2007); E. L. Quarantelli, "Disaster Planning, Emergency Management, and Civil Protection: The Historical Development of Organized Efforts to Plan for and to Respond to Disasters," University of Delaware Disaster Research Center, Preliminary Paper Series, no. 301 (2000), http://udspace.udel.edu/handle/19716/673.

12. While SEMS inspired NIMS, it also had to be adjusted after NIMS was adopted in order to be compatible.

13. "Informational releases are cleared through IC [Incident Command]/UC [Unified Command], the EOC [Emergency Operations Center]/MAC [Multiagency Coordination] Group, and/or Federal officials ... to ensure consistent messages, avoid release of conflicting information, and prevent negative impact on operations." *Foundation for the Standardized Emergency Management System*, public information section.

14. The disaster preparedness guidelines at all levels of government emphasize regional coordination between the multiple Joint Information Centers in order to create a cohesive Joint Information System.

15. *National Incident Management System*, 29.

16. Ibid., 51.

17. The process of getting information to the public includes several steps: gathering (from public information officers, people in "on-scene command," and the media), verifying, coordinating (including "establishing key messages"), and disseminating information "using multiple methods." Ibid., 73.

18. Like the national and state disaster plans, San Francisco's disaster plans include a number of annexes that are conceptually identical to the federal emergency support functions outlined in NIMS. The city emergency plan includes a number of emergency support function annexes that are to be used as needed during disasters. The roles and activities described in the annexes are to be activated if the scale or complexity of an emergency demands it. These annexes are primarily divided in ways that reflect the federal government's division of functions. The *San Francisco Emergency Response Plan ESF #15: Joint Information System Annex* is focused on the work involved in producing and monitoring public information. In contrast, the *San Francisco Emergency Response Plan ESF# 2: Communications Annex* is primarily concerned with the functioning of the physical communication infrastructures as well as ensuring communications among city and state employees. *ESF #2* names the wide variety of technologies involved in "public warning communications

systems": commercial pages (on commercial paging systems); Emergency Alert System; Emergency Digital Information Service; KalW @91.7; "New Media include[ing] media outlets such as blogging, websites, and social networking sites (Twitter, Facebook, MySpace)"; National Oceanic and Atmospheric Administration; Outdoor Public Warning System; RoamSecure Notifications (CCSF Alert and AlertSF); and San Francisco Government Television. *City and County of San Francisco Emergency Response Plan Emergency Support Function #2: Communications Annex* (San· Francisco: City and County of San Francisco Emergency Management Program, 2009), http://sfdem.org/plans-0.

In contrast, *ESF #15* is primarily concerned with the ways in which the government emergency response touches the public; the framework is for the purpose of producing "emergency public information in the City and County of San Francisco." *City and County of San Francisco Emergency Response Plan Emergency Support Function #15: Joint Information System Annex* (San Francisco: City and County of San Francisco Emergency Management Program, 2015), 3.

19. As with many of the other disaster plans from the modern era, *ESF #15* describes a complex set of organizational roles and relationships that shape the Joint Information System. There is a large hierarchical organization of people who are in charge of producing the messages that are to be distributed. The Emergency Operations Center's public information officer "is accountable to the EOC Manager who is ultimately responsible for ensuring that emergency public information conveys the proper tone and messaging priorities as directed by executive leadership." The *ESF #15* gives details about when the messages need to be reviewed and approved by the Emergency Operations Center manager. These include messages with obvious political implications: details about fatalities, "politically sensitive topics (e.g., FEMA denial of federal assistance requests)," and "messages disseminated in response to negative public sentiment." The Emergency Operations Center's public information officer can approve other messages. Ibid., 27.

ESF #15 describes how the Joint Information Section of the Emergency Operations Center will work along with the mayor's office, Incident Command System's public information officer, and rest of the Emergency Operations Center. For example, the plan lays out "continuous two-way communication" between the Joint Information Section and the "rest of the EOC" where the Joint Information Section shares "ground truth; situational awareness; requests for help; potential risks/threats; [and] updated messaging," and the Emergency Operations Center shares "verified facts/status updates; response priorities/efforts; info to invalidate rumors; potential public safety concerns; [and] subject matter expertise." The plan details how the coordination of different organizations works, even suggesting the frequency of calls and topics that the calls might cover. Ibid., 8.

20. Department of Homeland Security, Federal Emergency Management Agency Region IX, and California Governor's Office of Emergency Services, *California Cata-*

strophic Incident Base Plan: Concept of Operations, 2008, B-1, http://www.caloes.ca
.gov/for-individuals-families/catastrophic-planning.

21. Mica Endsley, "Toward a Theory of Situation Awareness in Dynamic Systems," *Human Factors* 37, no. 1 (1995): 32–64.

Situational awareness also migrated out of the military into the broader world of human factors research—a move that is often credited to work by Endsley in the early 1990s. Crisis informatics researchers today who seek to use Twitter to contribute to situational awareness have adopted Endsley's articulation of situational awareness as a cognitive process.

22. Governor's Office of Emergency Services et al., *San Francisco Bay Area Regional Emergency Coordination Plan*, 2008, 4–24, http://sfdem.org/plans-0.

In the city and county of San Francisco, disaster planning documents specifically note that social media are a source that may contribute to situational awareness. The Joint Information Section staff assists the Emergency Operations Center and other members of the government disaster response in "maintaining situational awareness, ... monitoring traditional and social media outlets, ... creating and disseminating messages." *San Francisco Emergency Response Plan Emergency Support Function #15*, 24.

Sometimes the disaster managers choose to only share these situational awareness reports privately, even though they may contain information generated by the public (such as first-person accounts of the event), and certainly have the potential to inform reporting to the public: "*situation reports (sitreps) and other situational updates ... for official use only (FOUO).*" Ibid., 22 (emphasis in original).

The current NIMS (published in 2008) uses the language of a *common operating picture*, meaning "an overview of an incident created by collating and gathering information—such as traffic, weather, actual damage, resource availability—of any type (voice, data, etc.) from agencies/organizations in order to support decision making." *National Incident Management System*, 23.

The new NIMS proposal specifically discusses the production of situational awareness through a complex sociotechnical assemblage that relies on information collected through social media, geographic information systems, radio, and telephone systems. *National Engagement Draft, National Incident Management System* (Washington, DC: US Department of Homeland Security, 2016), https://www.fema
.gov/media-library/assets/documents/118063.

23. *Emergency Support Function 15: Standard Operating Procedures* (Washington, DC: US Department of Homeland Security, 2013), R-3.

24. Ibid., R-4. "The team used social media for situational awareness including information about social media discussions on power outages, volunteering and donations, and sentiment about the response efforts ... [and this information] was shared with Department, ESF #15, and interagency leadership, as well as the

National Response Coordination Center, Joint Field Offices, and other important partners." Ibid.

25. Frances Fragos Townsend, *The Federal Response to Hurricane Katrina: Lessons Learned* (Washington, DC: Executive Office of the President, 2006), 41. Situational awareness is blamed for the poor federal response or offered as a solution multiple times throughout the document, such as this example: "The Secretary lacked real-time, accurate situational awareness of both the facts from the disaster area as well as the on-going response activities of the Federal, State, and local players." Ibid., 52. See also ibid., 36, 55, 57, 94.

26. *State of California Emergency Plan* (Sacramento: California Emergency Management Agency, 2009), 49. This definition is shared in other documents. For example, *Foundation for the Standardized Emergency Management System*; and *National Incident Management System.*

27. *State of California Emergency Plan*, 50.

28. The plans say that they are to be read and used by the "whole community"—a decidedly inclusive approach. The *National Protection Framework* is focused on assessing the "information sharing requirements" for all groups in the United States, with an interest in distributing information about "actionable measures" for protection. *National Protection Framework*, 1st ed. (Washington, DC: US Department of Homeland Security, 2014), 13. The *National Response Framework* emphasizes that "critical tasks" include informing the public about "critical," "protective," and "life-sustaining information." *National Response Framework*, 2nd ed. (Washington, DC: US Department of Homeland Security, 2013), 21.

29. *National Mitigation Framework*, 23.

30. *Emergency Support Function 15: Standard Operating Procedures*, A-11.

31. *San Francisco Emergency Response Plan Emergency Support Function #15*, 15, 23, 55, 14.

32. The Bay Area Emergency Public Information Network strategy document contains goals for improving public information and warning practices as well as an assessment of the available resources. The project was sponsored by the Urban Securities Initiative and funded by the Department of Homeland Security.

"San Francisco has a wide variety of means for disseminating warning messages to the public and has developed relatively robust protocols. These include Alert SF, a text-based message delivery program that delivers emergency information to cell phone and other text-enabled devices and email accounts. San Francisco also has 109 outdoor sirens located across the OA [Operational Area] designed to alert residents and visitors and has approximately 50,000 Twitter followers for disaster preparedness and response. While the OA can deliver messages in Cantonese and Spanish via sirens, there are remaining challenges for multi-lingual warnings to the variety of populations in the OA." *Bay Area Emergency Public and Information Warning Strategic Plan 2012–2017* (San Francisco: Bay Area Urban Areas Security Initiative, 2012), 8.

33. The *City and County of San Francisco Emergency Response Plan: Earthquake Annex* describes the damage a catastrophic earthquake could do, and walks through three different scenarios. "Sources for this information include studies performed by: San Francisco's Community Action Plan for Seismic Safety (CAPSS) Draft Report; The Association of Bay Area Governments (ABAG); Charles Kircher and Associates; and USGS." *City and County of San Francisco Emergency Response Plan: Earthquake Annex* (San Francisco: City and County of San Francisco Emergency Management Program, 2008), 70.

Scenario planning is a popular "anticipatory technology" in disaster preparedness. Ben Anderson, "Preemption, Precaution, Preparedness: Anticipatory Action and Future Geographies," *Progress in Human Geography* 34, no. 6 (December 1, 2010): 777–798; Peter Adey and Ben Anderson, "Anticipating Emergencies: Technologies of Preparedness and the Matter of Security," *Security Dialogue* 43, no. 2 (April 1, 2012): 99–117; Ben Anderson and Peter Adey, "Affect and Security: Exercising Emergency in 'UK Civil Contingencies,'" *Environment and Planning D: Society and Space* 29, no. 6 (2011): 1092–1109; Andrew Lakoff, "Preparing for the Next Emergency," *Public Culture* 19, no. 2 (April 1, 2007): 247–271.

34. *San Francisco Emergency Response Plan: Earthquake Annex*, 55. The multifaceted "response strategy" of the city includes developing standardized messages, prioritizing safety and wellness, coordinating with TV and radio, using AlertSF, and using newsletters and flyers "to communicate to people in shelters and other areas of congregation." Ibid., 56.

35. The *San Francisco Emergency Response Plan: Earthquake Annex* provides a list of the "critical earthquake situation information" elements that are needed and in what time period. Ibid., 21, 25.

36. "[The city and county of San Francisco] has activated the emergency response plan and is fully deploying all available resources. Please continue to monitor emergency broadcast radio information releases, remain calm and work together with your neighbors to support each other through these next difficult days. If you have neighbors with disabilities or other impairments, please check on their welfare and support them in any way possible." Ibid., 57.

37. The authority to do this kind of work—to push messages to people—is located in Executive Order 13407, which requires "an effective, reliable, integrated, flexible, and comprehensive system to alert and warn the American people." George W. Bush, "Public Alert and Warning System, Executive Order 13407" (2006), https://www.federalregister.gov/d/06-5829.

38. The Emergency Alert System started in 1997 and is "designed for the President to speak to the American people within 10 minutes." It is a successor to the Emergency Broadcast System, which ran from 1963 to 1997. FEMA traces the "Evolution of Public Emergency Alerting" as including CONELRAD, which was envisioned as a way to warn citizens of nuclear disaster. "The Evolution of Public Emergency Alerting" (Washigton, DC: Federal Emergency Management Agency, 2014), https://www.fema.gov/media-library/assets/documents/27183.

39. IPAWS integrates different channels, such as the Emergency Alert System, with the Commercial Mobile Alert System and Wireless Emergency Alerts as well as a National Oceanic and Atmospheric Administration program called HazCollect, which manages nonweather emergency messages for the National Weather Service. See the course "FEMA Emergency Management Institute (EMI) Course IS-247.A: Integrated Public Alert and Warning System (IPAWS)," *Federal Emergency Management Agency*, October 31, 2013, https://training.fema.gov/is/courseoverview.aspx?code=IS-247.a. See also "Integrated Public Alert and Warning System," *Federal Emergency Management Agency*, February 28, 2016, https://www.fema.gov/integrated-public-alert-warning-system.

40. The California Office of Emergency Services advertises that Wireless Emergency Alerts will not "get stuck in highly congested areas which can happen with standard mobile voice and texting services." California Office of Emergency Services, *Why Not WEA?*, 2015, https://youtu.be/Zmo8Mwzg_rw. However, in order for local organizations to use the Wireless Emergency Alerts, they must have IPAWS-compatible software to post their alerts. "Wireless Emergency Alerts: FAQs," *California Office of Emergency Services*, 2014, http://www.calalerts.org/FAQs.html.

41. "A 'system of systems' exists when a group of independently operating emergency public information and warning systems within each OA [Operational Area] comprised of people, organizations, plans, technology, and procedures are interoperable; enabling public health and safety personnel to effectively deliver emergency public information and warnings within and across jurisdictions in the Bay Area." The system of systems approach represents "a universal principal of multi-system interoperability that may be applied to the emergency public information and warning capability (and other capabilities)." *Bay Area Emergency Public and Information Warning Strategic Plan 2012–2017*, 23, 77.

42. "FEMA Emergency Management Institute (EMI) Course IS-247.A."

43. The CAP Alert Message format can be used to describe a message that can be "fed" into many different types of media delivery vehicles. CAP is a format developed by the OASIS technical standards body through a public process that invited comments from people all over the world.

44. The federal training asks IPAWS message producers to defer to their local rules around the issuance of alerts and also gives the following criteria: "1) Does the haz-

ardous situation require the public to take immediate action? Does the hazardous situation pose a serious threat to life or property? Is there a high degree of probability the hazard situation will occur?" "FEMA Emergency Management Institute (EMI) Course IS-247.A."

45. IPAWS training materials, however, warn that IPAWS does not provide translation services and caution against automatic translation, advising "you should validate any automatically translated text with a speaker of the language to avoid errors." Ibid.

46. Bruns and Burgess, "Crisis·Communication in Natural Disasters," 273.

47. See, for example, Leysia Palen, Starr Roxanne Hiltz, and Sophia B. Liu, "Online Forums Supporting Grassroots Participation in Emergency Preparedness and Response," *Communications of the ACM* 50, no. 3 (2007): 54–58; Leysia Palen, Sarah Vieweg, Sophia B. Liu, and Amanda Lee Hughes, "Crisis in a Networked World: Features of Computer-Mediated Communication in the April 16, 2007, Virginia Tech Event," *Social Science Computer Review* 27, no. 4 (November 1, 2009): 467–480; Kate Starbird, Leysia Palen, Amanda Lee Hughes, and Sarah Vieweg, "Chatter on the Red: What Hazards Threat Reveals about the Social Life of Microblogged Information," in *Proceedings of the 2010 ACM Conference on Computer Supported Cooperative Work* (New York: ACM, 2010), 241–250; Leysia Palen and Sarah Vieweg, "The Emergence of Online Widescale Interaction in Unexpected Events: Assistance, Alliance, and Retreat," in *Proceedings of the 2008 ACM Conference on Computer Supported Cooperative Work* (New York: ACM, 2008), 117–126; Joanne I. White and Leysia Palen, "Expertise in the Wired Wild West," in CSCW '15 Proceedings of the 18th ACM Conference on Computer Supported Cooperative Work and Social Computing (New York: ACM Press, 2015), 662–675.

48. "The public should follow the instructions of first responders and emergency management officials. For more information visit www.sf72.org, tune your radio to KCBS 740 AM or 106.9 FM, or follow @SF_Emergency on Twitter." San Francisco Department of Emergency Management, "Earthquake Press Release Template."

49. Department of Homeland Security, Federal Emergency Management Agency Region IX, and California Governor's Office of Emergency Services, *California Catastrophic Incident Base Plan*, R-2.

50. California Office of Emergency Services, "Cal OES Home," http://www.caloes .ca.gov.

51. Ibid.

52. One assessment of Bay Area counties and major municipalities said that while these local governments "recognized the value of providing and monitoring information via social media ... few ... have a social media policy and several are not

monitoring social media or using their social media accounts." *Bay Area Emergency Public and Information Warning Strategic Plan 2012–2017*, 44.

53. The frequent changes in the popularity of platforms is noted in the federal *Concept of Operations* plan; the document envisions that the "guidelines in this Annex should carry over to any digital communication channel." *Emergency Support Function 15: Standard Operating Procedures*, R-1.

54. San Francisco Department of Emergency Management, "SF72—Get Connected," http://www.sf72.org/connect.

55. Ibid. The site explicitly advocates that people use social media to consider what they will do after a disaster: "In the event of an emergency, we all turn to our existing tools and networks—many of which are digital. Think about your connections and the online communities you're already a part of, so you can share updates and information when something happens." The site suggests signing up for Nextdoor (a website that connects people with their neighbors) and Airbnb, a short-term housing rental platform, "so you are ready to find or share a place to stay if anything happens."

56. Products such as Geofeedia explicitly used social media to allow users—frequently local governments—to keep tabs on citizens on social media. Matthew Cagle, "Facebook, Instagram, and Twitter Provided Data Access for a Surveillance Product Marketed to Target Activists of Color," *American Civil Liberties Union Free Future Blog*, October 11, 2016, https://www.aclu.org/blog/free-future/facebook -instagram-and-twitter-provided-data-access-surveillance-product-marketed.

Since Geofeedia's work was exposed, its contracts that allowed it access to Facebook and Twitter data have been terminated.

57. Aleksandra Sarcevic et al., "Beacons of Hope in Decentralized Coordination: Learning from on-the-Ground Medical Twitterers during the 2010 Haiti Earthquake," in *Proceedings of the ACM 2012 Conference on Computer Supported Cooperative Work* (ACM, 2012), 47–56; Kate Starbird and Leysia Palen, "Working and Sustaining the Virtual Disaster Desk," in *Proceedings of the 2013 Conference on Computer Supported Cooperative Work* (ACM, 2013), 491–502; Kate Starbird and Leysia Palen, "Voluntweeters: Self-Organizing by Digital Volunteers in Times of Crisis," in *Proceedings of the SIGCHI Conference on Human Factors in Computing Systems* (ACM, 2011), 1071–1080; Leysia Palen et al., "Scale in a Data-Producing Organization: The Socio-Technical Evolution of OpenStreetMap in Response to Humanitarian Events," in *Proceedings of the 33rd Annual ACM Conference on Human Factors in Computing Systems*, CHI '15 (New York, NY, USA: ACM, 2015), 4113–4122.

58. For example, "we observe, evaluate, and visualize how Twitter data evolves over time before, during, and after a natural disaster such as Hurricane Sandy and what opportunities there may be to leverage social media for situational awareness and emergency response." Robert M. Patton, Chad A. Steed, Chris G. Stahl, and Jim N.

Treadwell, "Observing Community Resiliency in Social Media," in *Computational Science and Its Applications—ICCSA 2013*, ed. Beniamino Murgante, Sanjay Misra, Maurizio Carlini, Carmelo M. Torre, Hong-Quang Nguyen, David Taniar, Bernady O. Apduhan, and Osvaldo Gervasi (Heidelberg: Springer, 2013), 491–501. "This growing use of social media during crises offers new information sources from which the right authorities can enhance emergency situation awareness. … By leveraging the public's collective intelligence, emergency authorities could better understand 'the big picture' during critical situations, and thus make the best, most informed decisions possible for deploying aid, rescue and recovery operations" Jie Yin, Sarvnaz Karimi, Andrew Lampert, Mark Cameron, Bella Robinson, and Robert Power, "Using Social Media to Enhance Emergency Situation Awareness," *IEEE Intelligent Systems* 27, no. 6 (2012): 52–59. "This paper reports on development of a web-based geovisual analytics application, *SensePlace2*, through which information contained in social media can be gathered and analyzed to support situational awareness domains." Alan M. MacEachren et al, "SensePlace2: GeoTwitter Analytics Support for Situational Awareness," in *VAST 2011—IEEE Conference on Visual Analytics Science and Technology 2011, Proceedings* (New York: IEEE, 2011), 181–190. These university-based projects were funded by the Australian National Security Science and Technology Branch within the Department of the Prime Minister, US Department of Homeland Security, and Oak Ridge National Laboratory managed by UT-Battelle, LLC, for the US Department of Energy.

59. Mirca Madianou, Liezel Longboan, and Jonathan Corpus Ong, "Finding a Voice through Humanitarian Technologies? Communication Technologies and Participation in Disaster Recovery," *International Journal of Communication* 9 (2015): 3020–3038; Ryan Burns, "Rethinking Big Data in Digital Humanitarianism: Practices, Epistemologies, and Social Relations," *GeoJournal* 80, no. 4 (August 2015): 477–490; Ryan Burns, "Moments of Closure in the Knowledge Politics of Digital Humanitarianism," *Geoforum* 53 (May 2014): 51–62; Lilie Chouliaraki, "RE-MEDIATION, INTER-MEDIATION, TRANS-MEDIATION: The Cosmopolitan Trajectories of Convergent Journalism," *Journalism Studies* 14, no. 2 (April 2013): 267–283; Taylor Shelton, Ate Poorthuis, Mark Graham, and Matthew Zook, "Mapping the Data Shadows of Hurricane Sandy: Uncovering the Sociospatial Dimensions of 'Big Data,'" *Geoforum* 52 (March 2014): 167–179; Mirca Madianou, "Digital Inequality and Second-Order Disasters: Social Media in the Typhoon Haiyan Recovery," Social Media + Society 1, no. 2 (2015).

60. Shelton et al., "Mapping the Data Shadows of Hurricane Sandy"; Kate Crawford and Megan Finn, "The Limits of Crisis Data: Analytical and Ethical Challenges of Using Social and Mobile Data to Understand Disasters," *GeoJournal* 80, no. 4 (November 1, 2014): 1–12.

61. danah boyd and Kate Crawford, "Critical Questions for Big Data," *Information, Communication, and Society* 15, no. 5 (2012): 662–679.

62. Kevin Driscoll and Shawn Walker, "Big Data, Big Questions | Working within a Black Box: Transparency in the Collection and Production of Big Twitter Data," *International Journal of Communication* 8 (June 16, 2014): 1745–1764.

63. Lucy Suchman, "Situational Awareness: Deadly Bioconvergence at the Boundaries of Bodies and Machines," *MediaTropes* 5, no. 1 (2015).

64. As Lucy Suchman says,

While critics rightly direct our attention to questions regarding the ethical and legal status of mechanized decision-making, I am suggesting here that we focus on a prior question regarding the promise of "decision" itself. Arguably always a fictive prelude to action, the moment of decision becomes further distributed across messy assemblages of socio-technical mediation that presuppose the recognizability of their objects, at the same time that those objects become increasingly difficult to define.

Ibid, 19.

65. As James Beninger argues in *The Control Revolution*, information technology facilitates the consolidation of decision-making among a few people. James R. Beniger, *The Control Revolution: Technological and Economic Origins of the Information Society* (Cambridge, MA: Harvard University Press, 1986).

66. It should be noted that there are projects that explicitly seek to improve situational awareness for ordinary people, the "true 'first responders.'" Palen, Hiltz, and Liu, "Online Forums Supporting Grassroots Participation in Emergency Preparedness and Response."

67. Mark Zuckerberg, Facebook.com, October 15, 2014, https://www.facebook.com/zuck/posts/10101699265809491.

68. Naomi Gleit, Sharon Zeng, and Peter Cottle, "Introducing Safety Check," *Facebook Newsroom*, October 15, 2014, http://newsroom.fb.com/news/2014/10/introducing-safety-check.

69. Gillespie,"Relevance of Algorithms."

70. This gives it a sampling biased to frequent Facebook users. Peter Cottle and Brian Sa, "Mass Messaging: Reaching Millions during Large-Scale Disasters," September 14, 2015, https://youtu.be/ptsCWGZW_P8. A note on sources for Facebook's Safety Check algorithm: because this code is not open source, I had to rely on public talks given by Facebook developers, press releases, quasi-press releases by Facebook executives posted to Facebook.com, and the testimonies of journalists, bloggers, and others who shared their experience as part of Facebook's calculated disaster publics.

71. This approach led to "race conditions" (a term of art among coders that refers to when computations are happening at the same time and not in the proper sequence) where people might receive multiple Safety Check message requests—something that the developers thought was "not that bad." Ibid.

72. Ibid.

73. A lot of this kind of work was already happening using the Facebook platform. Josh Constantine, "Facebook Safety Check Now Lets Locals Find and Offer Community Help-Like Shelter," *TechCrunch.com*, February 8, 2017, https://techcrunch .com/2017/06/14/facebooks-safety-check-will-integrate-fundraisers-among-other-upgrades; Naomi Gleit, "Empowering People to Help One Another within Safety Check," *Facebook Newsroom*, February 8, 2017, https://newsroom.fb.com/news/2017/ 02/empowering-people-to-help-one-another-within-safety-check.

74. Facebook takes a processing fee of 6.9% plus 30 cents—after terrorist attacks in May in Manchester, UK, the company reported that over $450K was collected for those affected. Facebook's newer features also allow people to comment on their safety check-in and provide more "context" about a disaster from "trusted third party" companies. Sarah Perez, "Facebook's Safety Check Will Integrate Fundraisers, among Other Upgrades," *TechCrunch.com*, June 14, 2017, https://techcrunch. com/2017/06/14/facebooks-safety-check-will-integrate-fundraisers-among-other-upgrades; Naomi Gleit, "Announcing Updates to Safety Check," *Facebook Newsroom*, June 14, 2017, https://newsroom.fb.com/news/2017/06/announcing-updates-to-safety-check.

75. "Google Person Finder is one of the tools that the Google Crisis Response team uses. The Google Crisis Response team analyzes the scale of impact of the disaster and then determines which of its tools would be most useful for responding to the given situation." Google.org, "Person Finder Help," 2016, https://support.google .com/personfinder.

76. People who input records set the expiration date for when the record will be deleted; if they do not set an expiration date, Google.org will remove the database from online access when it deems it is appropriate to do so. Ibid.

77. The Google.org system includes data imported from other groups by applying for an application programming interface (API) key. As the Person Finder FAQ explains, "All data entered into Google Person Finder is available to the public and searchable and accessible by anyone. Google does not review or verify the accuracy of the data. The standard Google Terms of Service apply to all users of Google Person Finder." Ibid.

78. You can find information out about someone by listing their cell phone number or part of their name. Google.org returns the records for matches to any part of the name search string or the nickname, including address and status. If you have information about someone, you are presented with a form that asks you for a home address (presumably before the disaster) and leaves space for a message to this person. In 2012, Person Finder would return names that were parts of a matched a string, including even a middle name. The Google.org code of 2012 suggested that if "Jr." was not in the record of a person, and people put the name "Jr." in a search

string, they would not necessarily find the results of the person they were looking for. The Google.org Person Finder code is available as open source. Ibid.; Google.org, "Person Finder—2012 Philippines Floods," www.google.org/personfinder/2012-08-philippines-flood.

79. However, Google (the corporation) has returned a tailored page of search results for ongoing disasters that directs people to the Person Finder tool by Google.org.

80. James C. Scott, *Seeing Like a State: How Certain Schemes to Improve the Human Condition Have Failed* (New Haven, CT: Yale University Press, 1998).

81. If the point is to find, identify, and target particular types of people (based on demographic information, for example), these systems would facilitate this. "Thus, unless one wishes to make an ethical-philosophical case that no state ought to have such panoptic powers—and hereby commit oneself to foregoing both its advantages (e.g., the Center for Disease Control) as well as its menace (like fine-combed ethnic cleansing)—one is reduced to feeding Leviathan and hoping, perhaps through democratic institutions, to tame it." James C. Scott, John Tehranian, and Jeremy Mathias, "The Production of Legal Identities Proper to States: The Case of the Permanent Family Surname," *Comparative Studies in Society and History* 44, no. 1 (January 2002): 4–44.

82. Some Facebook users who marked themselves as "safe" when they weren't in Nepal were blamed for being disrespectful to the people in Nepal, rather than Facebook being blamed for the error. Simon McCormack, "People Are Abusing a Facebook Tool Meant to Help People in Nepal," *Huffington Post*, May 13, 2015, Technology sec., http://www.huffingtonpost.com/2015/05/13/facebook-safety-check -nepal_n_7275802.html; Warren Rossalyn, "People Living in Britain and America Keep Marking Themselves as 'Safe' from Nepal's Earthquakes," *BuzzFeed News*, May 13, 2015, http://www.buzzfeed.com/rossalynwarren/people-living-in-britain-and -america-keep-marking-themselves.

83. As the *Atlantic* reported, in the case of using Safety Check after the Nepal earthquake, "Six Napalese [*sic*] of seven are not registered on the social network." Matt Schiavenza, "Updating Facebook to Say 'I'm Safe,'" *Atlantic*, April 25, 2015, http:// www.theatlantic.com/international/archive/2015/04/telling-the-world-youre-safe -through-facebook/391484.

84. Many people noticed this, and Schultz posted a lengthy online reply on November 14, 2015, explaining some of the criteria that Facebook used to decide when to deploy the Safety Check product. Alex Schultz, "Facebook Safety," Facebook.com, November 14, 2015, https://www.facebook.com/fbsafety/posts/930229667014872.

85. Michael Watts, "On the Poverty of Theory: Natural Hazards Research in Context," in *Interpretations of Calamity from the Viewpoint of Human Ecology*, ed. Kenneth Hewitt (Boston: Allen and Unwin Inc., 1983); Anthony Oliver-Smith, "Anthropo-

logical Research on Hazards and Disasters," *Annual Review of Anthropology* 25 (1996): 303–328; Anthony Oliver-Smith, "Theorizing Disasters: Nature, Power, and Culture," in *Catastrophe and Culture: The Anthropology of Disaster*, ed. Susanna M. Hoffman and Anthony Oliver-Smith (Santa Fe, NM: School of American Research Press, 2002); Kathleen Tierney, "From the Margins to the Mainstream?: Disaster Research at the Crossroads," *Annual Review of Sociology* 33 (2007): 504–525.

86. Constantine, "Facebook Safety Check Now Lets Locals Find and Offer Community Help-Like Shelter"; Gleit, "Empowering People to Help One Another within Safety Check." Also see iJet, at https://www.ijet.com/who-we-are/company-profiles, and NC4, at http://nc4.com/Pages/CompanyOverview.aspx.

87. Cottle and Sa, "Mass Messaging"; Todd Hoff, "How Facebook's Safety Check Works," *High Scalability*, November 14, 2015, http://highscalability.com/blog/2015/11/14/how-facebooks-safety-check-works.html.

88. Cottle and Sa, "Mass Messaging."

89. "The Emergency Alert System (EAS) is a national public warning system that requires broadcasters, cable television systems, wireless cable systems, satellite digital audio radio service (SDARS) providers, and direct broadcast satellite (DBS) providers to provide the communications capability to the President to address the American public during a national emergency. The system also may be used by state and local authorities to deliver important emergency information, such as AMBER alerts and weather information targeted to specific areas." "Emergency Alert System (EAS)," *Federal Communications Commission*, December 6, 2010, https://www.fcc.gov/general/emergency-alert-system-eas.

90. Metz Cade, "How Facebook Is Transforming Disaster Response," *WIRED*, November 10, 2016, https://www.wired.com/2016/11/facebook-disaster-response.

91. The *National Mitigation Framework* sets the goal of creating a "risk-conscious culture." Yet, these frameworks see the government as the producer of information, and the public as the receiver of it. *National Mitigation Framework*, 1st ed. (Washington, DC: US Department of Homeland Security, 2013), 1.

Chapter 6

1. A question that was challenging to address through analysis of these earthquakes is: Would earthquake publics living far away from friends and families have had the same impulse to calm their families before the telegraph existed, knowing that their families might not hear of the earthquake for weeks? For related discussions, see Gareth Davies, "Dealing with Disaster: The Politics of Catastrophe in the United States, 1789–1861," *American Nineteenth Century History* 14, no. 1 (March 2013): 53–72; Michele Landis Dauber, *The Sympathetic State : Disaster Relief and the Origins of the American Welfare State* (Chicago: University of Chicago Press, 2013); Megan Finn,

"Information Infrastructure and Descriptions of the 1857 Fort Tejon Earthquake," *Information & Culture: A Journal of History* 48, no. 2 (May 11, 2013): 194–221.

2. For an example of online rumor-spreading, see Cynthia Andrews, Elodie Fichet, Stella Ding, Emma Spiro, and Kate Starbird, "Keeping Up with the Tweet-Dashians: The Impact of 'Official' Accounts on Online Rumoring." Presented at ACM 2016 Computer Supported Cooperative Work (CSCW 2016).

3. Tarleton Gillespie, "The Relevance of Algorithms," in *Media Technologies: Essays on Communication, Materiality, and Society*, ed. Tarleton Gillespie, Pablo J. Boczkowski, and Kirsten A. Foot (Cambridge, MA: MIT Press, 2014), 167–194.

4. Mirca Madianou, Liezel Longboan, and Jonathan Corpus Ong, "Finding a Voice through Humanitarian Technologies? Communication Technologies and Participation in Disaster Recovery," *International Journal of Communication* 9 (2015): 3020–3038; Mirca Madianou, "Digital Inequality and Second-Order Disasters: Social Media in the Typhoon Haiyan Recovery," *Social Media + Society* 1, no. 2 (2015); Taylor Shelton, Ate Poorthuis, Mark Graham, and Matthew Zook, "Mapping the Data Shadows of Hurricane Sandy: Uncovering the Sociospatial Dimensions of 'Big Data,'" *Geoforum* 52 (March 2014): 167–179.

5. Ryan Burns, "Rethinking Big Data in Digital Humanitarianism: Practices, Epistemologies, and Social Relations," *GeoJournal* 80, no. 4 (August 2015): 477–490; Madianou, Longboan, and Ong, "Finding a Voice through Humanitarian Technologies?"; Megan Finn and Elisa Oreglia, "A Fundamentally Confused Document: Situation Reports and the Work of Producing Humanitarian Information," in *Proceedings of the 19th ACM Conference on Computer-Supported Cooperative Work and Social Computing* (New York: ACM Press, 2016), 1349–1362.

6. Boris Jardine and Christopher M. Kelty, "Preface: The Total Archive," *Limn*, no. 6 (March 4, 2016), http://limn.it/preface-the-total-archive; Charles Taylor, *Modern Social Imaginaries* (Durham, NC: Duke University Press, 2004).

7. Donna Haraway, "Situated Knowledges: The Science Question in Feminism and the Privilege of Partial Perspective," *Feminist Studies* 14, no. 3 (1988): 575–599.

8. See, for example, Kate Starbird and Leysia Palen, "Working and Sustaining the Virtual Disaster Desk," in *Proceedings of the 2013 Conference on Computer Supported Cooperative Work* (New York: ACM Press, 2013), 491–502; Leysia Palen, Robert Soden, T. Jennings Anderson, and Mario Barrenechea, "Scale in a Data-Producing Organization: The Socio-Technical Evolution of OpenStreetMap in Response to Humanitarian Events," in *Proceedings of the 33rd Annual ACM Conference on Human Factors in Computing Systems* (New York: ACM Press, 2015), 4113–4122; Leysia Palen, Starr Roxanne Hiltz, and Sophia B. Liu, "Online Forums Supporting Grassroots Participation in Emergency Preparedness and Response," *Communications of the ACM* 50, no. 3 (2007): 54–58.

9. James C. Scott, *Seeing Like a State: How Certain Schemes to Improve the Human Condition Have Failed* (New Haven, CT: Yale University Press, 1998); JoAnne Yates, *Control through Communication: The Rise of System in American Management*, rep. ed. (Baltimore: Johns Hopkins University Press, 1993); James R. Beniger, *The Control Revolution: Technological and Economic Origins of the Information Society* (Cambridge, MA: Harvard University Press, 1986).

10. Lucy Suchman, "Situational Awareness: Deadly Bioconvergence at the Boundaries of Bodies and Machines," *MediaTropes* 5, no. 1 (2015): 1–24.

11. José van Dijck, *The Culture of Connectivity: A Critical History of Social Media* (Oxford: Oxford University Press, 2013).

12. Burns, "Rethinking Big Data in Digital Humanitarianism"; Ryan Burns, "Moments of Closure in the Knowledge Politics of Digital Humanitarianism," *Geoforum* 53 (May 2014): 51–62; Starbird and Palen, "Working and Sustaining the Virtual Disaster Desk"; Kate Starbird, "Delivering Patients to Sacré Coeur: Collective Intelligence in Digital Volunteer Communities," in *Proceedings of the SIGCHI Conference on Human Factors in Computing Systems* (New York: ACM Press, 2013), 801–810.

13. People using the social media tools contribute most of the content on these platforms; they are not paid by the platforms. And when newspaper companies produce content that is distributed on these platforms, it is the *platforms* that earn the advertising dollars. Companies and scholars can pay for access to some data, but most people can't access these data. Platform owners get to decide what to do with the data even if they didn't generate it. Tiziana Terranova, "Free Labor: Producing Culture for the Digital Economy," *Social Text* 18, no. 2 (2000): 33–58.

14. Naomi Klein, *The Shock Doctrine*: The Rise of Disaster Capitalism (New York: Metropolitan Books, 2007).

15. Paul du Gay, *In Praise of Bureaucracy: Weber, Organization, Ethics* (London: Sage Publications, 2000); Daniel Kreiss, Megan Finn, and Fred Turner, "The Limits of Peer Production: Some Reminders from Max Weber for the Network Society," *New Media & Society* 13, no. 2 (March 1, 2011): 243–59.

16. Megan Finn, Janaki Srinivasan, and Rajesh Veeraraghavan, "Seeing with Paper: Government Documents and Material Participation," in *2014 47th Hawaii International Conference on System Sciences* (Los Alamitos, CA: IEEE Computer Society, 2014), 1515–1524.

17. Michele Landis, "Let Me Next Time Be 'Tried by Fire': Disaster Relief and the Origins of the American Welfare State, 1789–1874," *Northwestern University Law Review* 92, no. 3 (1998): 967–1034; Dauber, *The Sympathetic State*; Gareth Davies, "Dealing with Disaster: The Politics of Catastrophe in the United States, 1789–1861," *American Nineteenth Century History* 14, no. 1 (March 2013): 53–72; Gareth Davies,

"The Emergence of a National Politics of Disaster, 1865–1900," *Journal of Policy History* 26, no. 3 (July 2014): 305–326.

18. Russell R. Dynes, "The Dialogue between Voltaire and Rousseau on the Lisbon Earthquake: The Emergence of a Social Science View," University of Delaware Disaster Research Center, Preliminary Paper Series, no. 293 (1999), http://udspace.udel.edu/handle/19716/435.

19. On emergency anticipatory governance techniques, see Peter Adey and Ben Anderson, "Anticipating Emergencies: Technologies of Preparedness and the Matter of Security," *Security Dialogue* 43, no. 2 (April 1, 2012): 99–117; Ben Anderson and Peter Adey, "Affect and Security: Exercising Emergency in 'UK Civil Contingencies,'" *Environment and Planning D: Society and Space* 29, no. 6 (2011): 1092–1109; Ben Anderson, "Preemption, Precaution, Preparedness: Anticipatory Action and Future Geographies," *Progress in Human Geography* 34, no. 6 (December 1, 2010): 777–798; Stephen J. Collier and Andrew Lakoff, "Vital Systems Security: Reflexive Biopolitics and the Government of Emergency," *Theory, Culture, and Society* 32, no. 2 (March 1, 2015): 19–51.

20. Ulrich Beck, *Risk Society: Toward a New Modernity*, trans. Mark Ritter (London: Sage Publications, 1992); Anthony Giddens, *The Consequences of Modernity* (Stanford, CA: Stanford University Press, 1990).

21. Lee Clarke, *Mission Improbable: Using Fantasy Documents to Tame Disaster* (Chicago: University of Chicago Press, 1999).

22. *Bay Area Emergency Public and Information Warning Strategic Plan 2012–2017* (San Francisco: Bay Area Urban Areas Security Initiative, 2012).

23. National Research Council, "Overview: Lessons and Recommendations from the Committee for the Symposium on Practical Lessons from the Loma Prieta Earthquake," in *Practical Lessons from the Loma Prieta Earthquake: Report from a Symposium Sponsored by the Geotechnical Board and the Board on Natural Disasters of the National Research Council* (Washington, DC: National Academies Press, 1994).

24. These ideals are enacted through legislation such as the 1950 Stafford Act.

25. The US government is surveilling its citizens in the interest of combating terrorism with the authorization of secret courts—hardly a transparent process. These rules were brought into place in two days, without typical legislative debate, less than two months after the World Trade Center bombings in 2001. Kate Tummarello, "Debunking the Patriot Act as It Turns 15," Electronic Frontier Foundation, October 26, 2016, https://www.eff.org/deeplinks/2016/10/debunking-patriot-act-it-turns-15.

26. Assuming there are benefits of a relaxation of privacy and information policy norms, some may advocate that enacting particular activities in an emergency situation requires that a disaster is a clearly delineated period. Yet forty years of social science research has challenged the very terms of emergency, crisis, and disaster as

discursive constructions that can be difficult to consistently bind in time. Most researchers agree that the conditions of emergency are rooted in the everyday. Moreover, the experience of crisis can be relative; if middle-class people were suddenly thrust into impoverished living conditions, this experience might feel like an ongoing, chronic emergency. Kathleen Tierney, "From the Margins to the Mainstream?: Disaster Research at the Crossroads," *Annual Review of Sociology* 33 (2007): 504–525; Scott Gabriel Knowles, *The Disaster Experts: Mastering Risk in Modern America* (Philadelphia: University of Pennsylvania Press, 2011); Michael Watts, "On the Poverty of Theory: Natural Hazards Research in Context," in *Interpretations of Calamity from the Viewpoint of Human Ecology*, ed. Kenneth Hewitt (Boston: Allen and Unwin Inc., 1983).

27. Collier and Lakoff, "Vital Systems Security"; Stephen J. Collier, "Enacting Catastrophe: Preparedness, Insurance, Budgetary Rationalization," *Economy and Society* 37, no. 2 (May 2008): 224–250; Stephen J. Collier and Andrew Lakoff, "Distributed Preparedness: The Spatial Logic of Domestic Security in the United States," *Environment and Planning D: Society and Space* 26, no. 1 (2008): 7–28; Stephen J. Collier and Andrew Lakoff, "The Vulnerability of Vital Systems: How 'Critical Infrastructure' Became a Security Problem," in *Securing "the Homeland": Critical Infrastructure, Risk, and (in)Security*, ed. Myriam Anna Dunn and Kristian Søby Kristensen (New York: Routledge, 2008): 17–39; Andrew Lakoff, "Preparing for the Next Emergency," *Public Culture* 19, no. 2 (April 1, 2007): 247–271.

28. Clarke, *Mission Improbable*; Megan Finn, "Information Infrastructure and Resilience in American Disaster Plans," in *The Sociotechnical Constitution of Resilience: A New Perspective in Managing Risk and Disaster*, ed. Sulfikar Amir (London: Palgrave Macmillan, 2018).

29. S. Hilgartner, "Overflow and Containment in the Aftermath of Disaster," *Social Studies of Science* 37, no. 1 (February 1, 2007): 153–158; Knowles, *Disaster Experts*; Scott Gabriel Knowles, "Learning from Disaster?: The History of Technology and the Future of Disaster Research," *Technology and Culture* 55, no. 4 (2014): 773–784; Kevin Rozario, *The Culture of Calamity: Disaster and the Making of Modern America* (Chicago: University of Chicago Press, 2007); Beck, *Risk Society*.

30. Collier and Lakoff, "Vulnerability of Vital Systems."

31. Tierney, "From the Margins to the Mainstream?"

32. Peter Adey, "Facing Airport Security: Affect, Biopolitics, and the Preemptive Securitisation of the Mobile Body," *Environment and Planning D: Society and Space* 27, no. 2 (April 2009): 274–295.

33. Giorgio Agamben, *State of Exception* (Chicago: University of Chicago Press, 2005).

34. Stephen Graham and Nigel Thrift, "Out of Order: Understanding Repair and Maintenance," *Theory, Culture, and Society* 24, no. 3 (May 1, 2007): 1–25.

Selected Bibliography

Adey, Peter. "Facing Airport Security: Affect, Biopolitics, and the Preemptive Securitisation of the Mobile Body." *Environment and Planning D: Society and Space* 27, no. 2 (April 2009): 274–295.

Adey, Peter, and Ben Anderson. "Anticipating Emergencies: Technologies of Preparedness and the Matter of Security." *Security Dialogue* 43, no. 2 (April 1, 2012): 99–117.

Agamben, Giorgio. *State of Exception*. Chicago: University of Chicago Press, 2005.

Aldrich, Michele L., Bruce A. Bolt, Alan E. Leviton, and Peter U. Rodda. "The 'Report' of the 1868 Haywards Earthquake." *Bulletin of the Seismological Society of America* 76, no. 1 (1986): 71–76.

Anand, Nikhil. "Leaky States: Water Audits, Ignorance, and the Politics of Infrastructure." *Public Culture* 27, no. 2 (January 1, 2015): 305–330.

Anand, Nikhil. "Municipal Disconnect: On Abject Water and Its Urban Infrastructures." *Ethnography* 13, no. 4 (December 1, 2012): 487–509.

Anand, Nikhil. "PRESSURE: The PoliTechnics of Water Supply in Mumbai." *Cultural Anthropology* 26, no. 4 (November 2011): 542–564.

Ananny, Mike. "Networked News Time: How Slow—or Fast—Do Publics Need News to Be?" *Digital Journalism* 4, no. 4 (May 18, 2016): 414–431.

Ananny, Mike. "Press-Public Collaboration as Infrastructure: Tracing News Organizations and Programming Publics in Application Programming Interfaces." *American Behavioral Scientist* 57, no. 5 (May 1, 2013): 623–642.

Anderson, Ben. "Preemption, Precaution, Preparedness: Anticipatory Action and Future Geographies." *Progress in Human Geography* 34, no. 6 (December 1, 2010): 777–798.

Anderson, Ben, and Peter Adey. "Affect and Security: Exercising Emergency in 'UK Civil Contingencies.'" *Environment and Planning D: Society and Space* 29, no. 6 (2011): 1092–1109.

Anderson, Benedict. *Imagined Communities: Reflections on the Origin and Spread of Nationalism*. Rev. ed. London: Verso Books, 2006.

Andrews, Cynthia, Elodie Fichet, Stella Ding, Emma Spiro, and Kate Starbird. "Keeping Up with the Tweet-Dashians: The Impact of 'Official' Accounts on Online Rumoring." Presented at ACM 2016 Computer Supported Cooperative Work (CSCW 2016).

Bayly, C. A. *Empire and Information: Intelligence Gathering and Social Communication in India, 1780–1870*. Cambridge: Cambridge University Press, 1996.

Beck, Ulrich. *Risk Society: Towards a New Modernity*. Translated by Mark Ritter. London: Sage Publications, 1992.

Beniger, James R. *The Control Revolution: Technological and Economic Origins of the Information Society*. Cambridge, MA: Harvard University Press, 1986.

Blanchette, Jean-François. "A Material History of Bits." *Journal of the American Society for Information Science and Technology* 62, no. 6 (2011): 1042–1057.

Boellstorff, Tom. "For Whom the Ontology Turns: Theorizing the Digital Real." *Current Anthropology* 57, no. 4 (August 2016): 387–407.

Borgman, Christine L. "The Invisible Library: Paradox of the Global Information Infrastructure." *Library Trends* 51, no. 4 (2003): 652–654.

Bowker, Geoffrey C. *Science on the Run: Information Management and Industrial Geophysics at Schlumberger, 1920–1940*. Cambridge, MA: MIT Press, 1994.

Bowker, Geoffrey C., Karen Baker, Florence Millerand, and David Ribes. "Toward Information Infrastructure Studies: Ways of Knowing in a Networked Environment." In *International Handbook of Internet Research*, edited by Jeremy Hunsinger, Lisbeth Klastrup, and Matthew Allen, 97–117. Dordrecht: Springer Netherlands, 2010.

Bowker, Geoffrey C., and Susan Leigh Star. *Sorting Things Out: Classification and Its Consequences*. Cambridge, MA: MIT Press, 1999.

Bowker, Geoffrey C., Stephan Timmermans, Adele E. Clark, and Ellen Balka, eds. *Boundary Objects and Beyond: Working with Leigh Star*. Cambridge, MA: MIT Press, 2015.

boyd, danah, and Kate Crawford. "Critical Questions for Big Data." *Information Communication, and Society* 15, no. 5 (2012): 662–679.

Brechin, Gray. *Imperial San Francisco: Urban Power, Earthly Ruin.* 2nd ed. Berkeley: University of California Press, 2006.

Bronson, William. *The Earth Shook, the Sky Burned: A Photographic Record of the 1906 San Francisco Earthquake and Fire.* San Francisco: Chronicle Books, 2006.

Brown, John Seely, and Paul Duguid. *The Social Life of Information.* Cambridge, MA: Harvard Business School Press, 2000.

Bruce, John R. *Gaudy Century: The Story of San Francisco's Hundred Years of Robust Journalism.* New York: Random House, 1948.

Bruns, Axel, and Jean Burgess. "Crisis Communication in Natural Disasters: The Queensland Floods and Christchurch Earthquakes." In *Twitter and Society,* edited by Katrin Weller, Axel Bruns, Jean Burgess, Merja Mahrt, and Cornelius Puschmann, 89:373–384. New York: Peter Lang, 2014.

Buckland, Michael K. "Information as Thing." *Journal of the American Society for Information Science* 42, no. 5 (1991): 351–360.

Buckland, Michael K. "What Is a 'Document'?" *Journal of the American Society for Information Science* 48, no. 9 (1997): 804–809.

Burns, Ryan. "Moments of Closure in the Knowledge Politics of Digital Humanitarianism." *Geoforum* 53 (May 2014): 51–62.

Burns, Ryan. "Rethinking Big Data in Digital Humanitarianism: Practices, Epistemologies, and Social Relations." *GeoJournal* 80, no. 4 (August 2015): 477–490.

Calhoun, Craig J. *Habermas and the Public Sphere.* Cambridge, MA: MIT Press, 1992.

Calhoun, Craig J. "The Idea of Emergency: Humanitarian Action and Global (Dis) Order." In *Contemporary States of Emergency: The Politics of Military and Humanitarian Interventions,* edited by Didier Fassin and Mariella Pandolfi, 29–58. Cambridge, MA: MIT Press, 2010.

Carse, Ashley. *Beyond the Big Ditch: Politics, Ecology, and Infrastructure at the Panama Canal.* Cambridge, MA: MIT Press, 2014.

Caughey, John Walton. "Hubert Howe Bancroft, Historian of Western America." *American Historical Review* 50, no. 3 (April 1945): 461.

Chandler, Robert J. "The California News-Telegraph Monopoly, 1860–1870." *Southern California Quarterly* 58, no. 4 (1976): 459–484.

Chouliaraki, Lilie. "RE-MEDIATION, INTER-MEDIATION, TRANS-MEDIATION: The Cosmopolitan Trajectories of Convergent Journalism." *Journalism Studies* 14, no. 2 (April 2013): 267–283.

Clarke, Lee Ben. *Mission Improbable: Using Fantasy Documents to Tame Disaster.* Chicago: University of Chicago Press, 1999.

Coen, Deborah R. *The Earthquake Observers*. Chicago: University of Chicago Press, 2013.

Collier, Stephen J. "Enacting Catastrophe: Preparedness, Insurance, Budgetary Rationalization." *Economy and Society* 37, no. 2 (May 2008): 224–250.

Collier, Stephen J., and Andrew Lakoff. "Distributed Preparedness: The Spatial Logic of Domestic Security in the United States." *Environment and Planning D: Society and Space* 26, no. 1 (2008): 7–28.

Collier, Stephen J., and Andrew Lakoff. "Vital Systems Security: Reflexive Biopolitics and the Government of Emergency." *Theory, Culture, and Society* 32, no. 2 (March 1, 2015): 19–51.

Collier, Stephen J., and Andrew Lakoff. "The Vulnerability of Vital Systems: How 'Critical Infrastructure' Became a Security Problem." In *Securing "the Homeland": Critical Infrastructure, Risk, and (in)Security*, edited by Myriam Anna Dunn and Kristian Søby Kristensen. New York: Routledge, 2008.

Comfort, Louise K. *Shared Risk: Complex Systems in Seismic Response*. New York: Pergamon Press, 1999.

Crawford, Kate, and Megan Finn. "The Limits of Crisis Data: Analytical and Ethical Challenges of Using Social and Mobile Data to Understand Disasters." *GeoJournal* 80, no. 4 (November 1, 2014): 1–12.

Cuaresma, Jesús Crespo, Jaroslava Hlouskova, and Michael Obersteiner. "Natural Disasters as Creative Destruction? Evidence from Developing Countries." *Economic Inquiry* 46, no. 2 (2008): 214–226.

Dauber, Michele Landis. *The Sympathetic State: Disaster Relief and the Origins of the American Welfare State*. Chicago: University of Chicago Press, 2013.

Davies, Andrea Rees. *Saving San Francisco: Relief and Recovery after the 1906 Disaster*. Philadelphia: Temple University Press, 2011.

Davies, Gareth. "Dealing with Disaster: The Politics of Catastrophe in the United States, 1789–1861." *American Nineteenth Century History* 14, no. 1 (March 2013): 53–72.

Davies, Gareth. "The Emergence of a National Politics of Disaster, 1865–1900." *Journal of Policy History* 26, no. 3 (July 2014): 305–326.

Davis, Tracy C. *Stages of Emergency: Cold War Nuclear Civil Defense*. Durham, NC: Duke University Press, 2007.

Day, Ronald E. *The Modern Invention of Information: Discourse, History, and Power*. Carbondale: Southern Illinois University Press, 2001.

Des Chene, Mary. "Locating the Past." In *Anthropological Locations: Boundaries and Grounds of a Field Science*, edited by Akhil Gupta and James Ferguson. Berkeley: University of California Press, 1997.

DiMaggio, Paul, and Walter W. Powell. "The Iron Cage Revisited: Collective Rationality and Institutional Isomorphism in Organizational Fields." *American Sociological Review* 48, no. 2 (1983): 147–160.

Downey, Gregory J. *Telegraph Messenger Boys: Labor, Technology, and Geography, 1850–1950*. New York: Routledge, 2002.

Driscoll, Kevin, and Shawn Walker. "Big Data, Big Questions | Working within a Black Box: Transparency in the Collection and Production of Big Twitter Data." *International Journal of Communication* 8 (June 16, 2014): 20.

du Gay, Paul. *In Praise of Bureaucracy: Weber, Organization, Ethics*. London: Sage Publications, 2000.

Edwards, Paul N. "Infrastructure and Modernity: Force, Time, and Social Organization in the History of Sociotechnical Systems." In *Modernity and Technology*, edited by Thomas J. Misa, Phillip Brey, and Andrew Feenberg. Cambridge, MA: MIT Press, 2003.

Edwards, Paul N. "Meteorology as Infrastructural Globalism." *Osiris* 21, no. 1 (2006): 229–250.

Edwards, Paul N. *A Vast Machine: Computer Models, Climate Data, and the Politics of Global Warming*. Cambridge, MA: MIT Press, 2010.

Edwards, Paul N. "Y2k: Millennial Reflections on Computers as Infrastructure." *History and Technology* 15, no. 1–2 (October 1998): 7–29.

Eisenstein, Elizabeth L. *The Printing Press as an Agent of Change: Communications and Transformations in Early-Modern Europe*. Vols. 1–2. New York: Cambridge University Press, 1979.

Endsley, Mica. "Toward a Theory of Situation Awareness in Dynamic Systems." *Human Factors* 37, no. 1 (1995): 32–64.

Ensmenger, Nathan. "Computers as Ethical Artifacts." *IEEE Annals of the History of Computing* 29, no. 3 (July 2007): 88–87.

Ethington, Philip J. *The Public City: The Political Construction of Urban Life in San Francisco, 1850–1900*. New York: Cambridge University Press, 1994.

Fassin, Didier, and Mariella Pandolfi, eds. *Contemporary States of Emergency: The Politics of Military and Humanitarian Interventions*. Cambridge, MA: MIT Press, 2010.

Faust, Drew Gilpin. *This Republic of Suffering: Death and the American Civil War*. New York: Knopf, 2008.

Finn, Megan. "Information Infrastructure and Descriptions of the 1857 Fort Tejon Earthquake." *Information and Culture: A Journal of History* 48, no. 2 (May 11, 2013): 194–221.

Finn, Megan. "Information Infrastructure and Resilience in American Disaster Plans." In *The Sociotechnical Constitution of Resilience: A New Perspective in Managing Risk and Disaster*, edited by Sulfikar Amir. London: Palgrave Macmillan, 2018.

Finn, Megan, and Elisa Oreglia. "A Fundamentally Confused Document: Situation Reports and the Work of Producing Humanitarian Information." In *Proceedings of the 19th ACM Conference on Computer-Supported Cooperative Work and Social Computing*, 1349–1362. New York: ACM Press, 2016.

Finn, Megan, Janaki Srinivasan, and Rajesh Veeraraghavan. "Seeing with Paper: Government Documents and Material Participation." In *2014 47th Hawaii International Conference on System Sciences*, 1515–1524. Los Alamitos, CA: IEEE Computer Society, 2014.

Fortun, Kim. *Advocacy after Bhopal: Environmentalism, Disaster, New Global Orders.* Chicago: University of Chicago Press, 2001.

Fowler, Dorothy H. *A Most Dreadful Earthquake: A First-Hand Account of the 1906 San Francisco Earthquake and Fire, with Glimpses into the Lives of the Phillips-Jones Letter Writers.* Oakland: California Genealogical Society, 2006.

Fradkin, Philip. *The Great Earthquake and Firestorms of 1906: How San Francisco Nearly Destroyed Itself.* Berkeley: University of California Press, 2005.

Fradkin, Philip. *Magnitude 8: Earthquakes and Life along the San Andreas Fault.* Berkeley: University of California Press, 1999.

Fraser, Nancy. "Rethinking the Public Sphere: A Contribution to the Critique of Actually Existing Democracy." *Social Text* 25–26 (1990): 56–80.

Fressoz, Jean-Baptiste. "Beck Back in the 19th Century: Towards a Genealogy of Risk Society." *History and Technology* 23, no. 4 (2007): 333–350.

Geschwind, Carl-Henry. *California Earthquakes: Science, Risk, and the Politics of Hazard Mitigation.* Baltimore: Johns Hopkins University Press, 2001.

Giddens, Anthony. *The Consequences of Modernity.* Stanford, CA: Stanford University Press, 1990.

Gillespie, Tarleton. "The Relevance of Algorithms." In *Media Technologies: Essays on Communication, Materiality, and Society*, edited by Tarleton Gillespie, Pablo J. Boczkowski, and Kirsten A. Foot, 167–194. Cambridge, MA: MIT Press, 2014.

Goltz, James D., and Dennis S. Mileti. "Public Response to a Catastrophic Southern California Earthquake: A Sociological Perspective." *Earthquake Spectra* 27, no. 2 (2011): 487–504.

Gordon, Mary McD., and Cameron King Jr. "Earthquake and Fire in San Francisco." *Huntington Library Quarterly* 48, no. 1 (1985): 69–79.

Graham, Stephen. "When Infrastructures Fail." In *Disrupted Cities: When Infrastructure Fails*, edited by Stephen Graham, 1–26. New York: Routledge, 2009.

Graham, Stephen, and Nigel Thrift. "Out of Order: Understanding Repair and Maintenance." *Theory, Culture, and Society* 24, no. 3 (May 1, 2007): 1–25.

Guggenheim, Michael. "Introduction: Disasters as Politics—Politics as Disasters." *Sociological Review* 62 (June 1, 2014): 1–16.

Habermas, Jürgen. *The Structural Transformation of the Public Sphere: An Inquiry into a Category of Bourgeois Society*. Translated by Thomas Burger. Cambridge, MA: MIT Press, 1991.

Hansen, Gladys, and Emmet Condon. *Denial of Disaster: The Untold Story and Photographs of the San Francisco Earthquake of 1906*. San Francisco: Cameron and Company, 1989.

Haraway, Donna. "Situated Knowledges: The Science Question in Feminism and the Privilege of Partial Perspective." *Feminist Studies* 14, no. 3 (1988): 575–599.

Henkin, David M. *The Postal Age: The Emergence of Modern Communications in Nineteenth-Century America*. Chicago: University of Chicago Press, 2006.

Hilgartner, S. "Overflow and Containment in the Aftermath of Disaster." *Social Studies of Science* 37, no. 1 (February 1, 2007): 153–158.

Hittell, Theodore Henry. *The California Academy of Sciences: A Narrative History, 1853–1906*. Edited by Alan E. Leviton and Michele L. Aldrich. San Francisco: California Academy of Sciences, 1997.

Jackson, Steven J. "Rethinking Repair." In *Media Technologies: Essays on Communication, Materiality, and Society*, edited by Tarleton Gillespie, Pablo J. Boczkowski, and Kirsten A. Foot, 221–240. Cambridge, MA: MIT Press, 2013.

Jackson, Steven J., Paul N. Edwards, Geoffrey C. Bowker, and Cory P. Knobel. "Understanding Infrastructure: History, Heuristics, and Cyberinfrastructure Policy." *First Monday* 12, no. 6 (2007): 1–9.

Jardine, Boris, and Christopher M. Kelty. "Preface: The Total Archive." *Limn*, no. 6 (March 4, 2016). http://limn.it/preface-the-total-archive.

John, Richard R. *Network Nation: Inventing American Telecommunications*. Cambridge, MA: Harvard University Press, 2010.

John, Richard R. *Spreading the News: The American Postal System from Franklin to Morse*. Cambridge, MA: Harvard University Press, 1995.

Jones, Marian Moser. *The American Red Cross from Clara Barton to the New Deal.* Baltimore: Johns Hopkins University Press, 2013.

Kaika, Maria, and Erik Swyngedouw. "Fetishizing the Modern City: The Phantasmagoria of Urban Technological Networks." *International Journal of Urban and Regional Research* 24, no. 1 (2000): 120–138.

Karasti, Helena, and Karen S. Baker. "Infrastructuring for the Long-Term: Ecological Information Management." In *Proceedings of the 37th Annual Hawaii International Conference on System Sciences, 2004.* IEEE, 2004.

Kelty, Christopher M. "Geeks, Social Imaginaries, and Recursive Publics." *Cultural Anthropology* 20, no. 2 (2005): 185–214.

Kelty, Christopher M. *Two Bits: The Cultural Significance of Free Software.* Durham, NC: Duke University Press, 2008.

Kielbowicz, Richard B. *News in the Mail: The Press, Post Office, and Public Information, 1700–1860s.* New York: Greenwood Press, 1989.

Klein, Naomi. *The Shock Doctrine: The Rise of Disaster Capitalism.* New York: Metropolitan Books, 2007.

Klinenberg, Eric. *Heat Wave: A Social Autopsy of Disaster in Chicago.* Chicago: University of Chicago Press, 2002.

Knowles, Scott Gabriel. *The Disaster Experts: Mastering Risk in Modern America.* Philadelphia: University of Pennsylvania Press, 2011.

Knowles, Scott Gabriel. "Learning from Disaster?: The History of Technology and the Future of Disaster Research." *Technology and Culture* 55, no. 4 (2014): 773–784.

Kuchinskaya, Olga. *The Politics of Invisibility: Public Knowledge about Radiation Health Effects after Chernobyl.* Cambridge, MA: MIT Press, 2014.

Lakoff, Andrew. "Preparing for the Next Emergency." *Public Culture* 19, no. 2 (April 1, 2007): 247–271.

Landis, Michele. "'Let Me Next Time Be 'Tried by Fire': Disaster Relief and the Origins of the American Welfare State, 1789–1874." *Northwestern University Law Review* 92, no. 3 (1998): 967–1034.

Larkin, Brian. "The Politics and Poetics of Infrastructure." *Annual Review of Anthropology* 42, no. 1 (October 21, 2013): 327–343.

Larkin, Brian. *Signal and Noise: Media, Infrastructure, and Urban Culture in Nigeria.* Durham, NC: Duke University Press, 2008.

Lave, Jean. *Apprenticeship in Critical Ethnographic Practice.* Chicago: University of Chicago Press, 2011.

Lave, Jean, and Etienne Wenger. *Situated Learning: Legitimate Peripheral Participation.* New York: Cambridge University Press, 1991.

Lawson, Andrew, G. K. Gilbert, Harry Fielding Reid, J. C. Branner, A. O. Leuschner, George Davidson, Charles Burkhalter, and W. W. Campbell. *The California Earthquake of April 18, 1906: Report of the State Earthquake Investigation Commission in Two Volumes and Atlas.* Washington, DC: Carnegie Institution of Washington, 1908.

Le Dantec, Chris. *Designing Publics.* Cambridge, MA: MIT Press, 2016.

Levy, David M. "Fixed or Fluid?: Document Stability and New Media." In *Proceedings of the 1994 ACM European Conference on Hypermedia Technology,* 24–31. New York: ACM Press, 1994.

Lund, Niels Windfeld. "Document Theory." *Annual Review of Information Science and Technology* 43, no. 1 (2009): 1–55.

Madianou, Mirca. "Digital Inequality and Second-Order Disasters: Social Media in the Typhoon Haiyan Recovery." *Social Media + Society* 1, no. 2 (2015).

Madianou, Mirca, Liezel Longboan, and Jonathan Corpus Ong. "Finding a Voice through Humanitarian Technologies? Communication Technologies and Participation in Disaster Recovery." *International Journal of Communication* 9 (2015): 3020–3038.

Marres, Noortje. "The Issues Deserve More Credit: Pragmatist Contributions to the Study of Public Involvement in Controversy." *Social Studies of Science* 37, no. 5 (October 1, 2007): 759–780.

Marres, Noortje. "Issues Spark a Public into Being: A Key but Often Forgotten Point of the Lippmann-Dewey Debate." In *Making Things Public: Atmospheres of Democracy,* edited by Bruno Latour and Peter Weibel, 208–217. Cambridge, MA: MIT Press, 2005.

Marres, Noortje. *Material Participation: Technology, the Environment, and Everyday Publics.* Houndmills, UK: Palgrave Macmillan, 2012.

Marres, Noortje. "Why Political Ontology Must Be Experimentalized: On Ecoshowhomes as Devices of Participation." *Social Studies of Science* 43, no. 3 (2013): 417–443.

Marvin, Simon, and Stephen Graham. *Splintering Urbanism: Networked Infrastructures, Technological Mobilities, and the Urban Condition.* London: Routledge, 2001.

Mindich, David T. Z. *Just the Facts: How "Objectivity" Came to Define American Journalism.* New York: NYU Press, 2000.

Nunberg, Geoffrey. "Farewell to the Information Age." In *The Future of the Book,* edited by Geoffrey Nunberg. 103–138. Berkeley: University of California Press, 1996.

Nye, David E. *American Technological Sublime*. Cambridge, MA: MIT Press, 1994.

Oakes, Guy. *The Imaginary War: Civil Defense and American Cold War Culture*. New York: Oxford University Press, 1994.

Odell, Kerry A., and Marc D. Weidenmeir. "Real Shock, Monetary Aftershock: The 1906 San Francisco Earthquake and the Panic of 1907." *Journal of Economic History* 64, no. 4 (December 2004): 1002–1027.

O'Donnell, Peggy. "Earthquakes: An Introduction." *Synergy* 1, no. 15 (March 1969): 1–6.

Oliver-Smith, Anthony. "Anthropological Research on Hazards and Disasters." *Annual Review of Anthropology* 25 (1996): 303–328.

Oliver-Smith, Anthony. "Theorizing Disasters: Nature, Power, and Culture." In *Catastrophe and Culture: The Anthropology of Disaster*, edited by Susanna M. Hoffman and Anthony Oliver-Smith. Santa Fe, NM: School of American Research Press, 2002.

Olson, Robert A., Shirley Mattingly, Charles Scawthorn, Jelena Pantelic, Dennis Mileti, Colleen Fitzpatrick, Steven Helmericks, et al. "Socioeconomic Impacts and Emergency Response." *Earthquake Spectra* 6, no. S1 (May 1, 1990): 393–431.

Ortner, Sherry B. "Theory in Anthropology since the Sixties." *Comparative Studies in Society and History* 26, no. 1 (1984): 126–166.

Owen, Graham. "After the Flood: Disaster Capitalism and the Symbolic Restructuring of Intellectual Space." *Culture and Organization* 17, no. 2 (March 2011): 123–137.

Palen, Leysia, Starr Roxanne Hiltz, and Sophia B. Liu. "Online Forums Supporting Grassroots Participation in Emergency Preparedness and Response." *Communications of the ACM* 50, no. 3 (2007): 54–58.

Palen, Leysia, Robert Soden, T. Jennings Anderson, and Mario Barrenechea. "Scale in a Data-Producing Organization: The Socio-Technical Evolution of OpenStreetMap in Response to Humanitarian Events." In *Proceedings of the 33rd Annual ACM Conference on Human Factors in Computing Systems*, 4113–4122. New York: ACM Press, 2015.

Palen, Leysia, and Sarah Vieweg. "The Emergence of Online Widescale Interaction in Unexpected Events: Assistance, Alliance, and Retreat." In *Proceedings of the 2008 ACM Conference on Computer Supported Cooperative Work*, 117–126. New York: ACM Press, 2008.

Palen, Leysia, Sarah Vieweg, Sophia B. Liu, and Amanda Lee Hughes. "Crisis in a Networked World: Features of Computer-Mediated Communication in the April 16, 2007, Virginia Tech Event." *Social Science Computer Review* 27, no. 4 (November 1, 2009): 467–480.

Pan, Erica Y. Z. *The Impact of the 1906 Earthquake on San Francisco's Chinatown*. New York: Peter Lang, 1995.

Parks, Lisa D., and Nicole Starosielski. *Signal Traffic: Critical Studies of Media Infrastructures*. Urbana: University of Illinois Press, 2015.

Prescott, William H. "Circumstances Surrounding the Preparation and Suppression of a Report on the 1868 California Earthquake." *Bulletin of the Seismological Society of America* 72, no. 6A (December 1, 1982): 2389–2393.

Reddy, Michael. "The Conduit Metaphor: A Case of Frame Conflict in Our Language about Language." In *Metaphor and Thought*, edited by Andrew Ortony, 284–324. Cambridge: Cambridge University Press, 1979.

Ribes, David, and Charlotte P. Lee. "Sociotechnical Studies of Cyberinfrastructure and E-Research: Current Themes and Future Trajectories." *Computer Supported Cooperative Work* 19, no. 3–4 (August 2010): 231–244.

Rozario, Kevin. *The Culture of Calamity: Disaster and the Making of Modern America*. Chicago: University of Chicago Press, 2007.

Sandvig, Christian. "The Internet as an Infrastructure." In *The Oxford Handbook of Internet Studies*, edited by William H. Dutton, 86–108. Oxford: Oxford University Press, 2013.

Sarcevic, Aleksandra, Leysia Palen, Joanne White, Kate Starbird, Mossaab Bagdouri, and Kenneth Anderson. "Beacons of Hope in Decentralized Coordination: Learning from On-the-Ground Medical Twitterers during the 2010 Haiti Earthquake." In *Proceedings of the ACM 2012 Conference on Computer Supported Cooperative Work*, 47–56. New York: ACM Press, 2012.

Schiller, Dan. *How to Think about Information*. Urbana: University of Illinois Press, 2007.

Schudson, Michael. "The Objectivity Norm in American Journalism." *Journalism* 2, no. 2 (August 1, 2001): 149–170.

Schumpeter, Joseph A. *Capitalism, Socialism, and Democracy*. 3rd ed. New York: Harper and Row, 1976.

Schwartz, Richard. *Earthquake Exodus, 1906: Berkeley Responds to the San Francisco Refugees*. Berkeley: RSB Books, 2005.

Schwarzlose, Richard. *The Nation's Newsbrokers*. Vol. 2, The Rush to Institution, from 1865 to 1920. Evanston, IL: Northwestern University Press, 1990.

Scott, James C. *Seeing Like a State: How Certain Schemes to Improve the Human Condition Have Failed*. New Haven, CT: Yale University Press, 1998.

Scott, James C., John Tehranian, and Jeremy Mathias. "The Production of Legal Identities Proper to States: The Case of the Permanent Family Surname." *Comparative Studies in Society and History* 44, no. 1 (January 2002): 4–44.

Shelton, Taylor, Ate Poorthuis, Mark Graham, and Matthew Zook. "Mapping the Data Shadows of Hurricane Sandy: Uncovering the Sociospatial Dimensions of 'Big Data.'" *Geoforum* 52 (March 2014): 167–179.

Sheriff, Patrick, ed. *2:46: Aftershocks: Stories from the Japan Earthquake*. London: Enhanced Editions Ltd., 2011.

Smith, Conrad. *Media and Apocalypse: News Coverage of the Yellowstone Forest Fires, Exxon Valdez Oil Spill, and Loma Prieta Earthquake*. Westport, CT: Greenwood Publishing Group, 1992.

Solnit, Rebecca. *A Paradise Built in Hell: The Extraordinary Communities That Arise in Disaster*. New York: Viking, 2009.

Solnit, Rebecca. *River of Shadows: Eadweard Muybridge and the Technological Wild West*. New York: Viking, 2003.

Srinivasan, Janaki, Megan Finn, and Morgan Ames. "Information Determinism: The Consequences of the Faith in Information." *Information Society* 33, no. 1 (2017): 13–22.

Star, Susan Leigh. "The Ethnography of Infrastructure." *American Behavioral Scientist* 43, no. 3 (November 1, 1999): 377–391.

Star, Susan Leigh. "Orphans of Infrastructure: A New Point of Departure." Paper presented at the Future of Computing: A Vision, Oxford Internet Institute, University of Oxford, March 2007.

Star, Susan Leigh, and Karen Ruhleder. "Steps toward an Ecology of Infrastructure: Design and Access for Large Information Spaces." *Information Systems Research* 7, no. 1 (1996): 111–134.

Starbird, Kate. "Delivering Patients to Sacré Coeur: Collective Intelligence in Digital Volunteer Communities." In *Proceedings of the SIGCHI Conference on Human Factors in Computing Systems*, 801–810. New York: ACM Press, 2013.

Starbird, Kate, and Leysia Palen. "Voluntweeters: Self-Organizing by Digital Volunteers in Times of Crisis." In *Proceedings of the SIGCHI Conference on Human Factors in Computing Systems*, 1071–1080. New York: ACM Press, 2011.

Starbird, Kate, and Leysia Palen. "Working and Sustaining the Virtual Disaster Desk." In *Proceedings of the 2013 Conference on Computer Supported Cooperative Work*, 491–502. New York: ACM Press, 2013.

Starbird, Kate, Leysia Palen, Amanda Lee Hughes, and Sarah Vieweg. "Chatter on the Red: What Hazards Threat Reveals about the Social Life of Microblogged Information." In *Proceedings of the 2010 ACM Conference on Computer Supported Cooperative Work*, 241–250. New York: ACM Press, 2010.

Stehle, Randy. "Auxiliary Markings: 'Burned Out' in the 1906 San Francisco Earthquake and Fire." *La Posta* 20, 6 (December 1989): 7–12.

Stehle, Randy. "Auxiliary Markings: 'Burned Out'; in the 1906 San Francisco Earthquake and Fire—Recent Discoveries and a Re-Examination of the Resumption of Normal Postal Service." *La Posta* 29, 3 (July 1998): 7–28.

Stehle, Randy. "The 1906 San Francisco Earthquake and Fire: Recent Discoveries (Part 3)." *La Posta* 33, no. 1 (March 2002): 47–50.

Stehle, Randy. *Postal History of the 1906 San Francisco Earthquake and Fire*. West Linn, OR: La Posta Publications, 2010.

Steinberg, Ted. *Acts of God: The Unnatural History of Natural Disaster in America*. 2nd ed. New York: Oxford University Press, 2006.

Steinberg, Ted. "Smoke and Mirrors: The San Francisco Earthquake and Seismic Denial." In *American Disasters*, edited by Steven Biel, 103–128. New York: NYU Press, 2001.

Subervi-Vélez, Federico, Maria Denney, Anthony Ozuna, Clara Quintero, and Juan-Vicente Palerm. *Communicating with California's Spanish-Speaking Populations: Assessing the Role of the Spanish-Language Broadcast Media and Selected Agencies in Providing Emergency Services*. Berkeley: California Policy Seminar, University of California at Berkeley, 1992.

Suchman, Lucy. "Situational Awareness: Deadly Bioconvergence at the Boundaries of Bodies and Machines." *MediaTropes* 5, no. 1 (2015): 1–24.

Taylor, Charles. *Modern Social Imaginaries*. Durham, NC: Duke University Press, 2004.

Terranova, Tiziana. "Free Labor: Producing Culture for the Digital Economy." *Social Text* 18, no. 2 (2000): 33–58.

Thomas, Gordon, and Max Morgan Witts. *The San Francisco Earthquake*. New York: Dell Publishing, 1971.

Tierney, Kathleen. "From the Margins to the Mainstream?: Disaster Research at the Crossroads." *Annual Review of Sociology* 33 (2007): 504–525.

Tierney, Kathleen. *The Social Roots of Risk: Producing Disasters, Promoting Resilience*. Stanford, CA: Stanford University Press, 2014.

Tierney, Kathleen, Christine Bevc, and Erica Kuligowski. "Metaphors Matter: Disaster Myths, Media Frames, and Their Consequences in Hurricane Katrina." *Annals of the American Academy of Political and Social Science* 604, no. 1 (March 2006): 57–81.

Tironi, Manuel. "Atmospheres of Indagation: Disasters and the Politics of Excessiveness." *Sociological Review* 62 (June 1, 2014): 114–134.

Tobriner, Stephen. *Bracing for Disaster: Earthquake-Resistant Architecture and Engineering, 1838–1933.* Berkeley, CA: Heyday Books, 2006.

Vaidhyanathan, Siva. "Afterword: Critical Information Studies." *Cultural Studies* 20, no. 2–3 (2006): 292–315.

van Dijck, José. *The Culture of Connectivity: A Critical History of Social Media.* Oxford: Oxford University Press, 2013.

Vaughan, Diane. *The Challenger Launch Decision: Risky Technology, Culture, and Deviance at NASA.* Chicago: University of Chicago Press, 1996.

Vaughan, Diane. "Theorizing Disaster: Analogy, Historical Ethnography, and the Challenger Accident." *Ethnography* 5, no. 3 (September 1, 2004): 315–347.

Wachtendorf, Tricia, and James M. Kendra. "Improvising Disaster in the City of Jazz: Organizational Response to Hurricane Katrina." In *Understanding Katrina: Perspectives from the Social Sciences.* Social Science Research Council, 2006.

Walker, Richard A. "California's Golden Road to Riches: Natural Resources and Regional Capitalism, 1848–1940." *Annals of the Association of American Geographers* 91, no. 1 (2001): 167–199.

Warner, Michael. "Publics and Counterpublics." *Public Culture* 14, no. 1 (2002): 49–90.

Watts, Michael. "On the Poverty of Theory: Natural Hazards Research in Context." In *Interpretations of Calamity from the Viewpoint of Human Ecology,* edited by Kenneth Hewitt. Boston: Allen and Unwin Inc., 1983.

Wermiel, Sara E. *The Fireproof Building: Technology and Public Safety in the Nineteenth-Century American City.* Baltimore: Johns Hopkins University Press, 2000.

White, Joanne I., and Leysia Palen. "Expertise in the Wired Wild West." In *CSCW '15 Proceedings of the 18th ACM Conference on Computer Supported Cooperative Work and Social Computing,* 662–675. New York: ACM Press, 2015.

Wollenberg, Charles. "Life on the Seismic Frontier: The Great San Francisco Earthquake (of 1868)." *California History* 71, no. 4 (Winter 1992–1993): 494–509.

Yates, JoAnne. *Control through Communication: The Rise of System in American Management.* Rep. ed. Baltimore: Johns Hopkins University Press, 1993.

Zornow, William Frank. "Jeptha H. Wade in California: Beginning the Transcontinental Telegraph." *California Historical Society Quarterly* 29, no. 4 (1950): 345–356.

Unpublished Research

Theses and Dissertations

Carter, John Denton. "The San Francisco Bulletin, 1855–1865: A Study in the Beginnings of Pacific Coast Journalism." PhD diss., University of California at Berkeley, 1941.

Laird, Ruth M. "Ethnography of a Disaster." Master's thesis, San Francisco State University, 1991.

Lemieux, Jessica. "Phoenix Rising: Effects of the 1906 Earthquake on California Print Culture." Master's thesis, San Jose State University, 2006.

Lovekamp, William E. "Gender, Race/Ethnicity, and Social Class Differences in Disaster Preparedness, Risk, and Recovery in Three Earthquake-Stricken Communities." PhD diss., Southern Illinois University at Carbondale, 2006.

Massey, Kimberly Kay. "A Qualitative Analysis of the Uses and Gratifications Concept of Audience Activity: An Examination of Media Consumption Diaries Maintained during the 1989 Loma Prieta Earthquake." PhD diss., University of Utah, 1993.

Rozario, Kevin. "Nature's Evil Dreams: Disaster and America, 1871–1906." PhD diss., Yale University, 1996.

Simile, Catherine M. "Disaster Settings and Mobilization for Contentious Collective Action: Case Studies of Hurricane Hugo and the Loma Prieta Earthquake." PhD diss., University of Delaware, 1995.

Sun, Yumei. "From Isolation to Participation: Chung Sai Yat Oo [*China West Daily*] and San Francisco's Chinatown, 1900–1920." PhD diss., University of Maryland at College Park, 1999.

Wachtendorf, Tricia. "Improvising 9/11: Organizational Improvisation following the World Trade Center Disaster." PhD diss., University of Delaware, 2004.

Reports and Professional Papers

Bolin, Robert C., and Lois M. Stanford. "Emergency Sheltering and Housing of Earthquake Victims: The Case of Santa Cruz County." In *The Loma Prieta, California, Earthquake of October 17, 1989—Public Response*, edited by Patricia Bolton, B43–50. US Geological Survey Professional Paper 1553-B. Washington, DC: US Government Printing Office, 1993.

Brown III, William M., and Carl Mortenson. "Earth Science, Earthquake Response, and Hazard Mitigation: Lessons from the Loma Prieta Earthquake." In *The Loma Prieta, California, Earthquake of October 17, 1989—Recovery, Mitigation, and*

Reconstruction, edited by Joanne M. Nigg, D81–90. US Geological Survey Professional Paper 1553-D. Washington, DC: US Government Printing Office, 1998.

Comerio, Mary C. "Hazard Mitigation and Housing Recovery: Watsonville and San Francisco One Year Later." In *The Loma Prieta, California, Earthquake of October 17, 1989—Recovery, Mitigation, and Reconstruction*, edited by Joanne M. Nigg, D29–34. US Geological Survey Professional Paper 1553-D. Washington, DC: US Government Printing Office, 1998.

Daggett, Emerson L., ed. *Trends in Size, Circulation, News, and Advertising in San Francisco Journalism*. Vol. 4. San Francisco: Works Progress Administration of Northern California, 1940.

Dynes, Russell R. "The Dialogue between Voltaire and Rousseau on the Lisbon Earthquake: The Emergence of a Social Science View." University of Delaware Disaster Research Center, Preliminary Paper Series, no. 293 (1999).

Edwards, Paul N., Steven J. Jackson, Geoffrey C. Bowker, and Cory P. Knobel. "Understanding Infrastructure: Dynamics, Tensions, and Design." Final report of the History and Theory of Infrastructure: Lessons for New Scientific Cyberinfrastructures workshop, Ann Arbor, MI, 2007.

Hansen, Gladys, and Frank R. Quinn. "The 1906 'Numbers' Game." Paper of the San Francisco Earthquake Research Project, San Francisco. 1985.

Phillips, Brenda. "Sheltering and Housing Low-Income and Minority Groups in Santa Cruz County after the Loma Prieta Earthquake." In *The Loma Prieta, California, Earthquake of October 17, 1989—Recovery, Mitigation, and Reconstruction*, edited by Joanne M. Nigg, D17–28. US Geological Survey Professional Paper 1553-D. Washington, DC: US Government Printing Office, 1998.

Phillips, Brenda, and Mindy Ephraim. "Living in the Aftermath: Blaming Processes in the Loma Prieta Earthquake." Natural Hazards Research Applications Information Center, Institute of Behavioral Science, University of Colorado, Natural Hazards Working Paper Series, no. 80 (September 1992).

Quarantelli, E. L. "Disaster Planning, Emergency Management, and Civil Protection: The Historical Development of Organized Efforts to Plan for and to Respond to Disasters." University of Delaware Disaster Research Center, Preliminary Paper Series, no. 301 (2000).

Quinn, Russell. *The San Francisco Press and the 1906 Fire*. Edited by Emerson L. Daggett. Vol. 5. San Francisco: Works Progress Administration of Northern California, 1940.

Rappaport, Richard J. "The Media: Radio, Television, and Newspapers." In *The Loma Prieta, California, Earthquake of October 17, 1989—Lifelines*, edited by Ansel J. Schiff,

A43–46. US Geological Survey Professional Paper 1552-A. Washington, DC: US Government Printing Office, 1998.

Rogers, Everett M., Matthew Berndt, John Harris, and John Minzer. "Accuracy in Mass Media Coverage." In *The Loma Prieta Earthquake: Studies of Short-Term Impacts*, edited by Robert C. Bolin, 75–83. Boulder: Institute of Behavioral Science, University of Colorado, 1990.

Schiff, Anshel J., Alex Tang, Lawrence F. Wong, and Luis Cusa. "Communication Systems." In *The Loma Prieta, California, Earthquake of October 17, 1989—Lifelines*, edited by Anshel J. Schiff, A24–25. US Geological Survey Professional Paper 1552-A. Washington, DC: US Government Printing Office, 1998.

Tierney, Kathleen. "Emergency Preparedness and Response." In *Practical Lessons from the Loma Prieta Earthquake: Report from a Symposium Sponsored by the Geotechnical Board and the Board on Natural Disasters of the National Research Council*, 105–134. Washington, DC: National Academies Press, 1994.

Tierney, Kathleen. "Emergency Medical Care Aspects of the Loma Prieta Earthquake." University of Delaware Disaster Research Center, Preliminary Paper Series, no. 161 (1991).

Index